*Feeling Time*

# *Feeling Time*

## Duration, the Novel, and Eighteenth-Century Sensibility

Amit S. Yahav

PENN

UNIVERSITY OF PENNSYLVANIA PRESS

PHILADELPHIA

Published by
University of Pennsylvania Press
Philadelphia, Pennsylvania 19104-4112
www.upenn.edu/pennpress

Printed in the United States of America on acid-free paper
1 3 5 7 9 10 8 6 4 2

Library of Congress Cataloging-in-Publication Data
Names: Yahav, Amit S., author.
Title: Feeling time : duration, the novel, and eighteenth-century sensibility / Amit S. Yahav.
Description: 1st edition. | Philadelphia : University of Pennsylvania Press, [2018] | Includes bibliographical references and index.
Identifiers: LCCN 2017047728 | ISBN 978-0-8122-5017-6 (hardcover : alk. paper)
Subjects: LCSH: English fiction—18th century—History and criticism. | Time in literature. | Time—Philosophy. | Time perception in literature. | Literature and society—England—History—18th century.
Classification: LCC PR830.T5 Y34 2018 | DDC 823/.509384—dc23
LC record available at https://lccn.loc.gov/2017047728

*To the luminous memory of Yossi*
*and to the vibrant presence of Ye'ela and Mikael*

# *Contents*

Introduction. The Sensibility Chronotope 1

Chapter 1. Composing Human Time: Locke, Hume, Addison, and Diderot 23

Chapter 2. Temporal Moralities and Momentums of Plot: Richardson and Hutcheson 50

Chapter 3. Sympathetic Moments and Rhythmic Narration: Sterne, Early Musicology, and the Elocutionists 82

Chapter 4. Durational Aesthetics and the Logic of Character: Radcliffe, Burke, and Smith 116

Coda. The End of Human Time? 148

Notes 153

Bibliography 183

Index 193

Acknowledgments 197

# Introduction

## The Sensibility Chronotope

SOON AFTER DISCOVERING a human footprint on his island, Robinson Crusoe concludes that "it must be some of the Savages of the main Land over-against me, who had wander'd out to Sea in their *Canoes* . . . [and] I should certainly have them come again in greater Numbers, and devour me" (113).[1] While he is right to suspect that the island's visitors are cannibals, he turns out to be wrong about the threat they pose to his life; the natives are unlikely to devour Crusoe upon encountering him, since they arrive at the island already equipped with all they need for their ritual, not in search of supplies for it. And just as they welcome as neighbors the survivors of the Spanish shipwreck (161), they are also likely to welcome Crusoe as a living friend rather than as dead foodstuff. Yet, if contrary to Crusoe's anxieties, other men do not eat and seem to have no intention of eating his body, they do consume his time.

While he thinks he is alone on the island, Crusoe approaches time as an abundant resource and an abstract measure; he enjoys a "prodigious deal of Time" (51), which he fills with a variety of tasks meticulously timed—twenty-four days to rescue supplies from his drowned ship (52), three and a half months to build a wall (56), two weeks for building a bower (75).[2] Indeed, during his initial years on the island, Crusoe feels he has "a World of Time" (79) at his disposal; "My Time or Labour was little worth, and so it was as well employ'd one way as another" (51), he confesses. But once he realizes that other humans are close by—from the moment he discovers the footprint on his island—his time no longer easily circulates among varying purposes, and he instead becomes solely devoted to formulating opinions about his new-found neighbors and devising strategies for an encounter. For "many Hours, Days; nay, I may say, Weeks and Months" (114), Crusoe is immobilized by anxiety, which then gives way to a spurt of defensive action—building a second fortification—and to superman fantasies: "For Night and Day, I could think of nothing but how I might destroy some of these Monsters in their cruel bloody Entertainment, and if possible, save the Victim they should bring hither to destroy" (122), which "pleas'd my Thoughts for some Weeks, and I was so full of it, that I often dream'd of it" (122). This self-aggrandizement then transforms into an effort at toleration, which lasts approximately another year (123), with Crusoe then sliding back to "above fifteen months . . . During all this Time, I was in the murthering Humour; and took up most of my Hours, which should have been better employ'd, in contriving how to circumvent, and fall upon them, the very next Time I should see them" (133). Then another two years of back and forth between vengeful superman fantasies and toleration, finally giving way to a pragmatic approach that leads Crusoe to a year and half's preparation for the opportunity "to get me a Servant, and perhaps a Companion" (146).

What might it mean that an anxiety for one's life and body materializes as an overwhelming of one's time? Or that the proximity of other people takes its toll in the form of an all-consuming duration? Or that one's sense of time comes to be indexed by alternations of mood? What kind of temporal conception supports such equivalences between bodies, life, feelings, and duration? And between one's relation to other people, on the one hand, and one's capacities for temporal command, on the other? Crusoe's conflation of the integrity of his body (his anxiety about cannibals) and the autonomy of his time (his reluctant absorption with his new-found neighbors) underlines a shift within the novel from an approach that takes time as an external resource, one that is especially abundant on the island and thus also circulates easily, to an approach that considers duration as endurance and links time with persons, thus not only impeding its circulation and contesting its abundance, but also endowing it with human, emotional, and embodied qualities. And Defoe's launching of this shift precisely at that point when Crusoe's supreme isolation no longer seems credible underlines how this turn is tied to a recognition of a shared world—that a profoundly human durational experience has much to do with a thoroughly social conception of existence.[3]

I begin with this brief sketch of temporal transformation in *Robinson Crusoe* as a gateway to the case that this book makes for a wider cultural shift toward identifying duration with human endurance and, as such, increasingly focusing on varying qualities of temporal experience. The cultural shift in temporal attitudes during the eighteenth century has usually been understood as the story of the development of a mechanical technology for counting time that came to pervade public life and individual consciousness. Influential histories have focused on rationalization, promoting a notion of modern temporal consciousness as governed by chronometry and geared to support the efficiency and power of the social totality at the price of thinning, or even fully draining out, durational qualities from personal and collective experience.[4] Programs of isolation and disciplinarity are, no doubt, key to eighteenth-century culture, as well as to modernity more generally. And yet eighteenth-century philosophy and literature have also undertaken extensive explorations of consciousness as a complex and nuanced interface of material, psychological, and social experience. Such investigations focus on the nexus of self and world, though not through frameworks of regimented schedules; and they underline a sociality different from, while also in complex relations with, the impersonal orders of commensurable exchange and print publicity. *Feeling Time* examines the vocabularies and logics used to explore temporal experience in such eighteenth-century discussions. It demonstrates that these yield accounts of duration that often attend to qualities no less than to quantities, intensities no less than extensities, and variations no less than regularities. It also finds that these eighteenth-century discussions identify felt duration as the crux of aesthetic pleasure and judgment, experiences described more as patterned durational activities than as static states.

In his analysis of "Duration and Its Simple Modes" in *An Essay Concerning Human Understanding*, John Locke famously argues that we derive our primary sense of duration from the succession of ideas, thus aligning his examination of time with his examination of the way our minds generate ideas from sensation and generate succession from what he calls "accidental Connexion."[5] Locke's account remained a constant touchstone for both academic and popular inquiries of time and of consciousness in England throughout the eighteenth century, with subsequent discussions elaborating on the sensible qualities of ideas and the compositional variations of succession. David Hartley, for example, develops Locke's associationism and sensationism in painstaking analyses of how various timings and orders of sensations—rather than their contents—yield meaningfulness. He identifies two forms of association, "the synchronous, and the successive," and argues that repetition of strings of sensation promotes memory and anticipation so that when a single sensation within the string is activated others in the string are also recollected.[6] Hartley thus suggests that a moment of sensation can be dilated into numerous associations, yielding analyses of near synchrony or, alternatively, compressed succession. And though Hartley does not comment on time directly, such recourse to memory and anticipation, synchrony and succession, makes it clear that regardless of what time might be in and of itself, it constitutes for him a fundamental operation of the mind.

Such eighteenth-century associationist and sensationist approaches resonate with more recent temporal phenomenologies in their insistent coupling of duration and experience—the double proposition that our sense of time arises from the operations of the mind and that the operations of the mind are constitutively temporal. Twentieth-century phenomenologies loop back the exploration of durational experience onto the question of what time is. For Edmund Husserl, the temporality of intention—the pattern by which a sensory present relates to what immediately precedes it (retention in Husserl's vocabulary) and follows it (protention)—promotes a notion of transcendental temporality—that time consciousness is the very structure that enables the differentiation of subject from object. For Martin Heidegger, Being's simultaneous groundedness in futurity (what Heidegger calls falling), pastness (existence), and presentness (facticity) becomes the fundamental support for an ontological idealism—the argument that ordinary time (external flow) is a form of originary temporality (the fundamentally temporalized structure of experience). And for Henri Bergson, recognizing states of consciousness as heterogeneous conglomerations of succession, near succession, and synchrony entails the notion of pure duration that refuses analogies of space and time, extensities and intensities, quantities and qualities. But eighteenth-century discussions do not extrapolate transcendental or ontological claims about time from temporal experience. Newton distinguishes between absolute time—time in and of itself—and relative time—our sense of duration—and Locke and Hume follow Newton in presupposing a strict separation between the two; relative time can at best approximate absolute time, but never be identical to it.[7] Moreover, Locke's and Hume's focus on what Newton calls relative time (though they use different terminology), involves a turn to what may best be described as habits: descriptions that refuse distinguishing minute levels of awareness—distinctions between pre-reflective and reflective consciousness. Yet without pressing explorations of durational experience back into a metaphysics, and while remaining at pragmatic levels of analysis, the eighteenth-century discussions I examine in *Feeling Time* offer rich analyses of felt duration, as well as of ethical and aesthetical implications of approaching time in this way. No less than Bergson's or Husserl's philosophies, many of the writings I examine consider how we perceive art as a salient analogy for how we feel time; no less than Bergson's or Heidegger's works, many of these discussions explore the temporal structures of authentic decision making and of care, and argue the case for these as durational experiences.[8]

My aim in this book, however, is not to recover a prehistory of twentieth-century phenomenology, but rather to consider eighteenth-century explorations of qualitative duration on their own terms and within the broad culture with which they are in conversation.[9] I begin this study with Locke's and Hume's comments on time, attending to their focus on mental processes that yield accounts of durational feelings. But these philosophies serve as points of departure for tracking engagement with qualitative durational experience across various genres. I consider Joseph Addison's and Denis Diderot's comments on the pleasures of reading, Francis Hutcheson's formulations of the moral sense, musicological and elocutionary treatises, Edmund Burke's and Adam Smith's aesthetic inquiries, and novels by Samuel Richardson, Laurence Sterne, and Ann Radcliffe. For all of these, temporality turns out to be key for psychoperceptual, ethical, and aesthetical explorations. Put together these discussions add up to what we might call a sensibility chronotope—shorthand for the temporal underpinnings of a culture that features a wide-ranging set of commitments to sensation, emotion, reflection, and sociability, and that develops alongside, though not in full agreement with, chronometric consciousness.

Sensibility, as many studies have shown, is an especially baggy and fluid category that conjoins feeling with thinking and judgment and is sometimes interchangeable with, and sometimes encompassing of, other terms such as sentiment, sentimentality, delicacy, and experience.[10] What began in the seventeenth century with research of psycho-perceptual processes—Newton's research of the vibratory constitution of nerve perception, Thomas Willis's explorations of the animal spirits, and Locke's analyses of the empirical origins of knowledge—came to be aligned in the eighteenth century with examinations of social relations and aesthetic preferences—Hutcheson's elaborations of a moral sense and a sense of beauty, Hume's and Smith's discussions of sympathy, as well as more popular treatises on domesticity, polite persuasion, and appreciation of art and literature.[11] Though not lending itself to the kinds of analyses that seek sharp focus and unequivocal differentiations, sensibility's capaciousness seems to me—as it seemed to late eighteenth- and early nineteenth-century writers—to be an especially useful concept. For sensibility's alignment of broad epistemological and moral concerns turned it into an encompassing worldview, a thoroughgoing culture that could house under the same title such varying manifestations as Richardson's didacticism and Sterne's irony, and whose precepts could extend beyond its orthodox proponents as well as beyond its heyday in the mid-eighteenth century.[12] Moreover, as Jane Austen makes clear through her compelling characterization of a Marianne Dashwood, sensibility's capaciousness enables it to stand for a wide spectrum of feelings—intuitive yet also reflective, strong yet also capable of composure, erroneous yet also thoroughly ethical, dynamic and varied yet also self-identical.[13] And sensibility's nuanced yet comprehensive approach to emotion well captures the range of feelings that *Feeling Time* discovers in eighteenth-century discussions of durational experience.

Underlying what I'm calling the "sensibility chronotope" is Mikhail Bakhtin's use of the term "chronotope" as a conjoining of culturally specific conceptions of time with their epistemological and moral underpinnings, as well as with their conventional forms of representation. If, as Giorgio Agamben claims, "Every culture is first and foremost a particular experience of time," then, according to Bakhtin, at stake is a complex feedback loop by which cultures, in their favoring of specific compositional forms over others, mediate temporality no less than they are constituted by it.[14] I should acknowledge, however, that the "chronotope" denotes spatiotemporal connectedness, and Bakhtin in the concluding remarks to his "Forms of Time and of the Chronotope in the Novel" stresses the inseparability of time and space. But throughout his influential study, Bakhtin treats time as "the dominant principle in the chronotope" (86)—a privileging already announced in the essay's title, with its double iteration of the temporal dimension.[15] In this study I follow Bakhtin's lead and focus on the temporal dimensions of the sensibility chronotope and do so for two reasons. First, we are still missing an extensive examination of sensibility's temporal underpinnings. Second, under the horizon of a modern chronometric consciousness, the connectedness of time and space has done much to obscure the rich and various dimensions of durational feelings.

James Chandler argues that at stake in sentimental representations is a logic of spatial, rather than temporal, transport that constitutes what he calls "sentimental probability."[16] Sentimental literature worries less about effects distant in time—the panning out of actions into their consequences—than about effects distant in space—how the conditions of any given moment are imagined to affect people occupying varying positions, not just characters, but also readers of novels and audiences of plays. Chandler demonstrates how from Shaftesbury's use of visual metaphors to underscore the reflective dimensions of soliloquy to Sterne's spatial and ocular descriptions of processes by which a spectator becomes spectacle, sentimentalists develop a unique logic of probability that relies on imagined swapping of positions located in space. This concern determines, in turn, a privileging of probable reflective and identificatory situations over probable plots. Chandler's privileging of the spatial over the temporal implicitly presumes a long tradition of scholarship that has focused on the eighteenth century's development of chronometry and chronology—an approach to time that, as Bergson recognized, spatializes duration and has only limited resources for recognizing varieties of temporal feelings. Bakhtin also contributes to such spatialization of time, albeit indirectly. When in his wide survey of chronotopes Bakhtin arrives at the eighteenth century, he identifies a form whose spatial components are key and whose temporal components are especially ghostly. Bakhtin claims that "a new feeling for time was beginning to awake" (228) in the eighteenth century, one that he calls nostalgia for the idyll. Nostalgia for the idyll associates emotion with recollection and pastness, sometimes also with hopefulness for future recuperation, but not with present experience. It is also a chronotope that strongly attaches feeling to a starkly differentiated spatial locale—to rural settings or to an insulated domestic sphere.[17] Bakhtin's discussion indicates that if in the eighteenth century a rising chronometric culture comes to pervade public life as well as private experience, then nostalgia for the idyll offers an alternative, but only in transport to a different time—past or future—or place—remote peripheries or enclosures. Nostalgia for the idyll thus highlights a compensatory imagination that coheres well with an understanding of the eighteenth century as the moment when temporal consciousness becomes impoverished by the rising power of an alienating rationalized approach. To feel the richness of idyllic time, one must be necessarily less than satisfied with one's present and imagine oneself elsewhere.

The "new feeling for time" that Bakhtin identifies in the eighteenth century, then, amounts to something like a negativity—a hollowing out of present temporal experience, compensated by imaginative transport along historical and geographical axes. But Stuart Sherman more recently demonstrates that chronometric consciousness need not be understood as a negativity. Sherman examines diurnal form as the textual counterpart of the invention of the minute and second hands on clocks and of the ratification of Greenwich Mean Time and longitude lines. By his analysis, these representational techniques enable precision and regularization while also securing opportunities for individualized content. "The particular forms of time proffered by the clocks, watches, and memorandum books so new and conspicuous in the period," he explains, "seemed to many serial autobiographers to limn a new temporality—of durations closely calibrated, newly and increasingly synchronized, and systematically numbered—durations that might serve as 'blanks' in which each person might inscribe a sequence of individual actions in an individual style" (18).[18] And yet even as chronometric notation increases opportunities for individualization, the timekeepers' paradigm gauges temporal experience by way of numbered measure and conceives of collective consciousness by way of abstractions. Thus there is little qualitative variation to an account of the duration of a walk that only imparts its length, to recall Sherman's evocative discussion of "minutes" in Pepys's diary (89), or to an account of the duration of work that only "refigures task time as tasks timed" (229), to recall his interpretation of *Robinson Crusoe*. Likewise, there is little social bonding through concurrent participation in abstract grids; in the final analysis, Sherman explains, such abstractions encourage an "obsessively cultivated privacy" (114) and promote "a larger program of textual isolation" (245).[19] And finally, conceiving of eighteenth-century temporal conceptions solely in terms of chronometric consciousness renders continuous the private and the public spheres, leisure and work time, subsuming all under the logic of utilitarian efficiency and abstract commensurability.

Addressing these concerns, Deidre Lynch considers the many ways in which eighteenth-century practices of leisure reading emphasized familial rituals of communal activity whose express purpose was to generate affective durations. Underwriting the valuation of these rituals was an associationist psychology that "identified the essence of feeling with its reiterative practice" (171), as Lynch puts it, and that operated on the primary levels of perceptual processes shared by all humans, as well as on the secondary processes of consciousness that constitute the cultural bonds of more specific communities. Nonetheless, Lynch assimilates this rich experience of leisure time to the chronometric logic that increasingly came to govern work and discipline. Building on Sherman's study, she presents literature's contribution to what she calls "quality time" as arising from the shift into a temporality of measure, and she highlights sensibility's steadying of emotion through practices that make reading seem like clockwork, emphasizing the extent to which these habits were supported by and promoted rigid schedules and diurnal form.[20]

Sherman and Lynch make nuanced cases for literature's participation in a chronometric culture, and they conceive of chronometric culture as complexly integrating feeling and individuation into the predominance of measure and standardization.[21] But while these studies have done much to complicate our understanding of what we might call a chronometric chronotope, in *Feeling Time* I delineate a sensibility chronotope that cannot be fully understood—or even perceived—from a perspective that presumes the primacy of chronometry and chronology. Alongside diurnal form and the persistence of idyll in the face of chronometry's ascendancy, we find in eighteenth-century literature a sensibility chronotope that, I will soon argue, might best be understood as a modern refiguration of romance. At stake in identifying a sensibility chronotope is recognizing the ways in which the literature of the period offers occasions for off-the-clock breaks as presence—in the present of reading—and as integral to modernity. And at stake in making the case for the sensibility chronotope as the refiguration of romance is to offer a way of understanding how this temporal conception is importantly tied—though not limited to—the genre of the novel.

William Wordsworth famously charges modern regularization with producing personalities afflicted with something like manic-depressive swings—"the increasing accumulation of men in cities, where the uniformity of their occupations produces a craving for extraordinary incident, which the rapid communication of intelligence hourly gratifies" and that also reduces minds "to a state of almost savage torpor."[22] Romantic poetry aims to ameliorate this condition of manic craving followed by depressive torpor by offering for its readers occasions for pausing from the ordinary business of "getting and spending," as Wordsworth puts it in his sonnet "The World Is Too Much with Us." It conceives of these lyrical pauses as tranquil recollections of passionate feeling, as Wordsworth would have it, or as intense insight being teased out of thought, as John Keats would have it. Thus Romantic poetry, as both reading material and a mode of thinking, offers its modern readers the opportunity to recompose into more varied and nuanced emotion and thinking. In its rural settings and its fascination with nature, such poetry's alternative to the chronometric resonates with Bakhtin's nostalgia for the idyll. And yet what distinguishes it from the chronotope that Bakhtin explores is its explicit thematization of acts of poetic thinking as off-the-clock breaks in the present for its speakers and readers. Nostalgia for the idyll cannot transform chronometric consciousness and only offers its readers occasions to recall other possibilities located at other times and places. Romantic poetry, by contrast, offers itself as an alternative that can be realized in the here and now of poetic reading and thinking, even as such alternate durational experiences can materialize only as temporary breaks.

Through such studies as M. H. Abrams's *Natural Supernaturalism* and Michael Clune's *Writing Against Time*, we have come to identify the possibility of pauses from our predominantly chronometric consciousness with the Romantic lyric and to take the alternative it offers as aiming for an atemporal eternity. Abrams maintains that the Romantics render significant moments as the "intersection of eternity with time" (385) and cast "timelessness as a quality of the experiential moment" (386).[23] More recently, but in important ways similarly, Clune describes the "Romantic quest to defeat time" (17) as focused on art's impossible ambition to sustain the intense sensory experience of first encounters.[24] And yet I argue in this book that by recognizing a sensibility chronotope, we can identify the ways in which literature has offered itself as pauses from the chronometric that are thoroughly durational, rather than atemporal. Indeed, Christopher Miller and Julia Carlson in studies that attend more carefully to the ways in which Romanticism draws on earlier eighteenth-century sources have been able to recognize poetry's mobilization of pacing and rhythm in techniques aimed to shape both represented durations and the duration of reading.[25] I show that a comprehensive sensibility chronotope has already in the eighteenth century self-consciously and fully conceived of leisure reading as off-the-clock durational breaks, and that prose writing and especially novels have most thoroughly taken up this task. That novels featured and provided such pauses for alternative duration is a capacity of the genre that eighteenth- and nineteenth-century readers and authors were well aware of, but that more recent scholarship has by and large overlooked. It may be that the Romantic lyric's more direct and programmatic response to the effects of rationalization has contributed to obscuring the ways in which an earlier sensibility approached this very same task. And yet the novel's orientation toward alternative durations arises precisely from what Romantic lyric and early novels share: a self-conscious engagement with, and refiguration of, romance.

If in the last quarter of the eighteenth century "the romance revival had become a major scholarly and poetic industry," as Ian Duncan puts it, then novels of sensibility have been preparing the grounds for such revival throughout the century.[26] From the perspective of novels of sensibility, romance isn't "revived" by Romanticism—neither by romantic lyric nor by the romantic novel—nor are Scott's historical novels the first to recognize romance as a mode of historicization. What Duncan identifies as the insight Scott discovers and develops from Gothic romances is only one way by which romance historicizes: delivering itself, as Duncan recognizes, "as modern culture's construction of a symbolic form prior to itself" (10–11). Romanticism continues early novels' persistent refiguration of romance, which involves not only—or even primarily—chronological priority, but a commitment to probable temporal experiences whose logic is other than chronological or chronometric.

Novelistic realism emerges as part and parcel of a renewed interest in historicism—as narratives that aim to tell the truth in a particular form, one that, as Elizabeth Ermarth emphasizes, constitutes empiricist epistemology not only as a commitment to descriptive detail but also as the prizing of a serial continuity that conceives of identity by way of cross-temporal comparisons. Medieval historiography, Ermarth reminds, relied on "the contrast between time and eternity and not upon the contrast between past and present"; the latter—which is the premise of a modern historiography that emerges in the Renaissance—"homogenized the temporal medium by finding past and present mutually informative" (25).[27] And, Ermarth continues, such "interest in the revelations of sequence . . . finds its fullest *aesthetic* expression in the temporal medium of literature and its fullest *literary* expression in the realistic novel, where the unfolding of structure and significance receives its most thoroughgoing serial treatment" (41). But as Michael McKeon argues, if what we take to be novelistic realism emerges from the epistemic transformations of "the early modern historicist revolution" (53), then what immediately precedes novels and their realism never ceases to leave its traces in them.[28] As McKeon points out, the form that precedes and persists most importantly in realist novels is romance. Romance has a remarkably long and diverse history, yet this genre endures, McKeon argues, because its most fundamental task is to figure an encounter of a nonhistoricist, nonchronometric mindset with a historicism that, more recently, we have come to identify as modern. Romance poses an alternative to such historicism precisely as it is embedded within it. McKeon's brief discussion of the Greek enlightenment demonstrates how romance succeeds myth. Romance thus delivers an encounter with this enlightenment's historical consciousness, rewriting mythic timelessness as "historical rupture between movement 'forward' and movement 'back' " (32). And McKeon's analysis of the twelfth-century Renaissance identifies romance as an encounter with historicism that continues to insist on a "qualitative standard of completeness" (38). If historicism pushes toward serial continuity, romance pushes toward an intuitive completeness that arises from human connectedness. "To be 'true' to another is a mode not of empirical veracity but of human connection coordinated by suprahuman principal [*sic*] or essence," McKeon explains (38).

Ermarth proposes that genres succeed one another discretely—from medieval revelational history to modern empiricist chronology, from typological plots to realistic plots—while allowing transitional phases, which she identifies with eighteenth-century novelists such as Defoe and Richardson. McKeon argues that genres internalize one another as they progress—a historiography that reshapes myth, a novel that refigures romance—and thus his analysis opens up room for an investigation of the refiguration of romance within modern chronometric consciousness and its allied novel form. *The Origins of the English Novel* does precisely that, but McKeon focuses on the components of response that we might call ideology—how debates about questions of truth and virtue came to be clinched together, thus transmuting and incorporating the plots of romance into the plots of realism to support a liberal regime that can at once house conservatism and progressivism. However, we can find in novels much more varied and nuanced techniques—well beyond emplotments and the specific ideologies these promote—for transmuting the temporal logic of romance. As I demonstrate extensively in subsequent chapters, eighteenth-century novelists developed techniques ranging from sentence-level effects of rhythm to patterns of memory and anticipation foregrounding time-consciousness as the bare bones of realist character; they also carefully attended to plot lines, not only for their causal chains, chronologies, and for endings' emplotments of meaning, but also for their complex constructions of metonymic relations between scenes and narrative wholes. All these techniques recast the principles of romance so as to feature qualitative durational experiences alongside chronometric frameworks and as part and parcel of a modern probabilistic culture.

Because romance is such a capacious form with such a long history, its principles are many. For the purpose of this study I rely on those identified by

Bakhtin, McKeon, and Patricia Parker, whose studies take into account the form's transhistorical persistence. In his analyses of the romance chronotope in Greek fiction and in chivalric tales, Bakhtin highlights two main principles. First, he identifies adventure time that functions as "a hiatus that leaves no *trace* in the life of the heroes or in their personalities" (90); thus "all the events of the novel that fill this hiatus are a pure digression from the normal course of life" (91). Second, he identifies distortions of time—"hours are dragged out, days are compressed into moments" (154). In her study of romance as a poetic mode with a long history from Ariosto to Keats, Parker elaborates the epistemological underpinnings of digressions; such dilations and suspensions of narrative progress, Parker argues, complexly and ambivalently explore errors as these impede a quest for truth. And McKeon, as we've already seen, identifies romance as founded on intuitive completeness through human connectedness underwritten by a suprahuman order.

All of these principles feature prominently in the sensibility chronotope, but they are also crucially revised by it. Adventures, rather than leaving no trace on the biographical time of characters, are grafted onto characters' bodies and minds. Everything Crusoe does and that happens to him after he stumbles across the footprint participates in his deliberations about the cannibals; it is manifest in his thinking, in mood swings, and in somatic responses such as nausea and vomit. Similarly, as we will see in subsequent chapters, Richardson insists on one's accounting for all adventures in one's life story, Sterne casts adventures as contests of opinion and then further attaches these opinions to pulsating bodies, and Radcliffe flags the key difference between gothic-romance characters and sensible-novelistic ones through the latter's integration of their adventures into continuous psychologies. Moreover, temporal distortions, rather than left standing as features of narration or plot, are presented as the qualitative variances of personal experience—the intensities of reading, conversation, thinking, and writing that can make any mere hour seem like an age not only in Sterne's renderings of conversations but also in Addison's discussions of leisure reading, in Richardson's lengthy presentations of moments, and in Radcliffe's version of the gothic as the supernatural explained. The epistemological quests that underline romance's digressions in Parker's account are refigured in sensibility as phenomenological descriptions. Instead of highlighting errors on a quest for truth, dilations in the sensibility chronotope draw attention to what such suspended moments feel like—what it feels like to imagine an unpredictable future in Richardson, or to talk not for the sake of getting things right but for the sake of staying in touch in Sterne, or to labor to craft consistency through whatever corpses or dolls, ghosts or nuns one encounters at any given moment in Radcliffe. Finally, the human connectedness that underwrites romance's sense of completion according to McKeon is revised by sensibility to punctuate the plot, rather than necessarily to end it, and to be explained not through suprahuman principles but through interpersonal durational experience. As we saw in the example from *Robinson Crusoe* and as we will see in Sterne's presentation of sympathy and in Radcliffe's presentation of courtship, connecting to other humans takes time, and this time is marked not as providential or unnatural intervention but, rather, as mood fluctuations of sympathy, anxiety, and intention.

One aim of this book, then, is to demonstrate how the sensibility chronotope refigures the temporal dimensions of romance through various techniques of plot, narration, and characterization to present qualitative duration within novels' diegetic worlds. But in arguing that the sensibility chronotope takes up a qualitative duration of off-the-clock breaks, I mean not only that within their fictions novels feature conceptions of qualitative temporality, but also that they aim to mediate such durational experiences for their readers. In attributing such a task of mediation to novels, I follow the lead of Paul Ricoeur, to whose seminal study *Time and Narrative* I want now to turn. Ricoeur has developed an especially cogent explanation of why and how narrative mediates temporal experience. He argues that Aristotle's mimesis encompasses not just the form of tragedy, but also, far more generally, the way narrative composition conjoins the temporality we see in action with the temporality we experience in thinking. He thus discovers in Aristotle's mimesis three different stages and locations of representational activity that are also constitutively temporal. What he calls *mimesis1*, or prefigured time, invokes symbolic structures of meaning; what he calls *mimesis2*, or configured time, refers to emplotted events; and what he calls *mimesis3*, or refigured time, denotes the reading process.[29] With such an expansion of mimesis, Ricoeur can establish that the referential significance of narrative representation has less to do with its better or worse mirroring of objects than with its activation of temporal experience constitutive of our actions and understanding. He can also establish that novels extend compositional principles to the representation of consciousness. Such expansion of action into consciousness challenges those who associate the novel with formlessness because of the genre's psychological emphases; the focus on minds, Ricoeur contends, highlights the very compositional dimensions that minds and narratives share. He thus reconceives our understanding of the referential function of representation to privilege mediation over imitation. He also calibrates our view of novelistic technique so that we can recognize compositional dimensions not only in the ordering of plot but also in characterization. And, finally, the focus on mediation and on composition enables Ricoeur to highlight the constitutive durationality of narrative, a duration that has often been obscured by formalist analyses.[30]

Ricoeur, however, understands experiential duration as fundamentally intellectual. His case studies of *Mrs. Dalloway*, *The Magic Mountain*, and *Remembrance of Things Past* privilege novels that feature characters and narrators directly contemplating time, thus risking conflating what he calls, following A. A. Mendilow, "tales about time" with philosophizing.[31] Moreover, though Ricoeur uses the word "experience" often in his discussions of reading, here too he focuses on the intellectual horizon of meaning, overlooking the meaningfulness that registers not just in understanding, but also in emotional and embodied effects. He thus cannot but underestimate eighteenth-century novels that focus their efforts less on communicating facts than on communicating "perception[s] of impressions," as a defender of sensibility in a 1796 installment of the *Monthly Magazine* puts it.[32] At stake is a literature that, as this eighteenth-century apologist continues, presumes that "no man is happy because he knows a truth, or believes a fact, but because he is conscious of a pleasing emotion" (708), and that aims less to inform its readers than to move them—to get them "sighing over a pathetic story, or weeping at a deep-wrought tragedy" (706). Think, for example, of *Tristram Shandy*'s relentless disavowals of its meaning and its displays of expressive bodies, and you might begin to imagine how the transformations of temporal experience that this novel famously achieves are never clarified by its notorious double philosophizing about time, but become more coherent in the somatic and emotional responses it evokes.[33] In order to gauge the durational dimensions of novels of sensibility I extend Ricoeur's analyses of literary mediation beyond concepts and understanding to attend to the material aspects of language and its effects and to the somatic and emotional registers of temporal consciousness. Sterne's and Radcliffe's prose, I suggest, cannot be fully accounted for without attention to its sonority; and Burke's aesthetics, I argue, cannot be fully understood without attention to his haptic descriptions of perception.

Additionally, my analyses of compositional construction extend beyond the elements of emplotment to which Ricoeur remains committed. Ricoeur focuses on the Aristotelian template of tragedy with its emphasis on causality, content, and the ethico-juridical perspective. He thus recoups characterological dimensions for compositional analysis, but emplotment remains the main paradigm in his tool kit. As he puts it in the introductory comments to his discussion of configured time in fiction (what he calls mimesis2): at stake is plot, defined as "an integrating dynamism that draws a unified and complete story from a variety of incidents" (2: 8). But the plot structure of novels of sensibility tends to be episodic—privileging incompleteness and noncausal relations—and these novels tend to offer multiple perspectives on the same events—highlighting variety in attitudes rather than variety in incidents.[34] But these novels suggest that the durational experience triggered by compositional activity—that refigured time activated by the configured time of representation—requires less causal logic (or its inverse in chance) than something more akin to musical patterning, which privileges similarity and contiguity. Thus in this study I emphasize how eighteenth-century associationist and sensationist philosophy highlights the contiguous and analogous compositional principles of consciousness, and how novels develop such noncausal compositional models in their techniques of narration and characterization.

The first chapter of this book considers the associationist and sensationist logics of eighteenth-century empiricism and their bearings on discussions of time. I begin by examining how Locke and Hume not only attend to our knowing time, but also identify duration with a consciousness that feels time and, more specifically, whose feelings of duration and about duration vary by intensities and compositional arrangement. If Gilles Deleuze reads Hume's account of human nature as a fully developed temporal phenomenology, in this chapter I discover the conditions of possibility for Hume's phenomenological approach in Locke's analysis of human understanding. I then turn to considering how these models of a temporalized consciousness are developed in Addison's and Diderot's comments on the durational makeups and effects of discursive compositions. Addison and Diderot emphasize that the compositional dimensions of durational experience and the qualities that these dimensions generate are made—that in preferring certain forms to others, writers shape the temporal experience of readers and can cultivate particular temporal qualities. Addison comments on relations between discursive compositions and time in a number of his *Spectator* pieces, presenting leisure reading as a means for evoking varying feelings by way of varying compositional principles. Such shaping of durational experience, according to Addison, constitutes the aesthetic pleasure of literature. Diderot brings this approach to bear on novels. According to Diderot, the pleasures of Richardson's novels arise from their combining conversational immediacy with lengthy narratives—a peculiar digressiveness that promotes attention to durations that we often overlook in our ordinary rush toward worldly achievements.

After this initial examination of eighteenth-century philosophy and popular essays, each of the subsequent chapters looks at a specific formal dimension of novels and considers the way it features the sensibility chronotope. Techniques of plot serve to develop solutions for difficulties of judging in time—the challenging assessment of action in medias res, while action continues to roll on to its destination and collide with other actions that divert it from its intended course—as well as to give readers occasions for experiencing the durationality of judgment. Techniques of narration serve to raise solutions for the difficulties of emotional attachments, helping to sustain love and sympathy for others through long durations by way of shared rhythms. And techniques of characterization engage the difficulty of bundling disjointed experiences into continuous identity, drawing on aesthetic theories that highlight the crafting of holistic durations out of disparate momentary stimulation and offering instances of aesthetic immersion as occasions for cultivating such continuous subjectivities.

Chapter 2 focuses on two related problems: the peculiar logic of plot in novels of sensibility and the difficulty of judging in time as it has been raised in theories of moral sentiments. While Smith's *The Theory of Moral Sentiments* may suggest, as Vivasvan Soni notes, that "if actions cannot reliably reach their *ends*, then the value of the ethical must be found elsewhere than in end-directed action," other models of sentimental morality continue to insist on what we might call a durational ethic that tackles action and its temporal discontents, as well as the experience of trying to comprehend such arcs from the position of an ever-rolling present.[35] Hutcheson and Richardson, I argue, develop a complex relation between the moment of experience and the continuums of narrative, suggesting that only such a double focus can yield an adequate understanding of what it means to be a moral agent not only operating within time but also constituted by time. Both writers value immediacy and authenticity—herein lies the significance of Hutcheson's "moral sense" and of Richardson's renderings of consciousness in flux—but both also insist on unpacking the moment as a position from which a fuller trajectory of the impelling force of action ought to be assessed. Hence Hutcheson commits to utilitarian calculations of the long arc and wide impact of actions, and Richardson insists on the necessary prolixity of his novels. Moral sentiments are thus incompletely surveyed when we focus on the spatial-spectatorial models of sympathetic identification while overlooking the durational models of narrative judgment. And Richardson's "writing to the moment" is misunderstood when presented as solely about effecting immediacy of character and presence, for it considers discrete segments of time not as freestanding experiences but as the very problem of configuring such segments into plots. With its insistence on both prolixity and immediacy, I further indicate, Richardson's technique refigures romance digression from an epistemological quest for truth into a phenomenological exploration of what it feels like to try to get things right when you know that you cannot possibly achieve such a result.

In Chapter 3 I turn to models of sympathy that privilege aurality and take up musical paradigms for conceptualizing both duration and community. The problem addressed here is the synchronization and endurance of fellow-feeling, a difficulty for which early musicology, eighteenth-century elocutionary theories, and novels of sensibility find solutions in rhythmic sound-strings. An understanding of sympathy as harmonious and rhythmic synchronization may blunt the passions and contrast with libertine renderings of love as excess, but it also enables emotion to be reliably shared over time. And it specifies sympathy as an occasion for experiencing a peculiar duration—one that, while relying on a recursive beat, is variously sonorous and emotional. The chapter looks at the early musicology of Roger North, Joshua Steele's importation of musical time into the study of elocution, and Laurence Sterne's sonorous narration as it combines with his notorious experiments in novelistic temporality. Sterne, I suggest, highlights the extent to which novels can convey the rhythms of language and thus at once represent moments of sympathetic immersion among characters and mediate such moments for readers. Sterne refigures both romance digression and its privileging of human connectedness as the fellow-feeling generated by sonorous discourse. Equally important, he considers how such sonorous and embodied sympathy might be communicated in print—made available at any time and any place. If literature consciously provides for ways in which its claims can be successfully iterated along long histories, as McKeon has argued, then for Sterne a novel's successful iteration in homogenous empty time relies on its ability to recreate particular qualities of duration—its capacity to mediate experience such that its pattern or order cannot be abstracted from human bodies that manifest and feel it.[36]

Chapters 2 and 3 engage theories of moral sentiments—looking, first, at their conceptions of the narrativity of action and judgment and, second, at their conceptions of the rhythmicality of sympathy. The final chapter of *Feeling Time* returns to aesthetic theories for their examinations of durational subjectivities. I draw attention to a tradition that extends Addison's durational aesthetics of reading to visual and sonorous artifacts. Burke emphasizes the temporality of sensation in *A Philosophical Enquiry into the Origin of Our Ideas of the Sublime and Beautiful*, and Smith highlights the compositional pleasure we take in instrumental music in his "Of the Nature of That Imitation Which Takes Place in What Are Called the Imitative Arts." Radcliffe adapts such insights for the purpose of thinking through the gothic motif of the eclipse of time as horror. Radcliffe, following Horace Walpole, identifies the genre of her fictions as romance; yet she also casts them as stories of sensibility. Thus while her plots are full of chance ruptures and her narration highlights what may be taken as distorted temporal perceptions—the end of time in romance's encounter with historicist chronometric consciousness—these principles are reshaped by sensibility into psychological processes. Radcliffe, I argue, devises a technique of characterization that relies not on details of content but on musical compositional form as theorized by Smith and on the processual nature of sensation as theorized by Burke. And she takes her mimetic reference to be the human mind as eighteenth-century associationist and sensationist philosophy conceives of it, highlighting the fundamental activity of crafting durational identities out of disparate moments. If realist characters are charged with representing the humanity of persons, then for Radcliffe this entails not underlining their individuality—as many scholars have presumed she ought to—but, rather, highlighting the typicality of the indispensable durational crafting that constitutes human consciousness.

During the last decades of the eighteenth century, gothic fiction viewed the eclipse of time as a possibility it needed to sound the alarm on vigorously. Today the alarm sounds no less urgently, warning us of an early twenty-first-century culture in which, as Jonathan Crary puts it, "the imposition of a machinic model of duration and efficiency onto the human body" (3) and the "generalized inscription of human life into duration without breaks" (8) effect "a static redundancy that disavows its relation to the rhythmic and periodic textures of human life" (8–9).[37] This is a 24/7 ideology, to which Crary provocatively ascribes horrific techniques of torture and military fantasies, as well as the more familiar but only slightly less disturbing daily realities of nonstop digital interfaces and consumerist opportunities. In Crary's account, it seems, romance has been so thoroughly internalized by realism that the historicist chronometric consciousness has come to seem like a suprahuman and hermetic principle. Crary clearly intends to be a provocateur, amplifying the familiar into the uncanny as a way of moving us to recognize—as gothic novels aimed to do—the grave risks our ordinary lives currently run. And while he is aware that twentieth- and twenty-first-century technologies only intensify and more fully realize the logic of capital and of disciplinarity launched some three hundred years ago, he also suggests that the kind of breaks for reconnecting with the various rich rhythms of durational experience that sensibility found in aesthetic pleasure, fellow-feeling, and judging have become increasingly rare. With the advent of televised entertainment and screen interfaces that are available almost anywhere and any time and that keep us connected to markets and to work, with the almost complete commodification of nature into remote and specially designated preserves, with an extremely capitalized art market that makes access to aesthetic pleasure rare and expensive, with friendships often conducted through screens and electronic messaging, and with judgment often reduced to consumer choices—the opportunities for pauses and our capacities for using them to recover thick and various human rhythms have significantly diminished. Moreover, if it was the very act of reading that the culture of sensibility believed could activate these capacities, then today we spend less time reading elaborate discursive strings, and fewer of us spend time in this way. Our exposure to a compositional experience of the world is diminishing and since, as Ricoeur reminds us, we still "have no idea of what a culture would be where no one any longer knew what it meant to narrate things" (2: 28), this is an especially frightening prospect.

"And yet . . . and yet" (2: 28), Ricoeur insists, and I agree. For Radcliffe and her contemporaries, occasions for transforming the chronometric into qualitative, profoundly human durations also seemed to demand—and to be worth—a struggle. It is not only that the apocalypse has been developing over at least three centuries, as Crary notes, but also that it has been recognized as the more specific threat of an end to human time for most of this period. We seem never to tire of telling stories about our end, and perhaps with such relentless narration of the threats to our endurance, we defer the end and enhance the qualities of the time that remains.[38] That we are an age of crisis is part and parcel of the modern sense-making form, Frank Kermode reminds us.[39] But it may also be that the machines, just like Defoe's cannibals, do not seek to annihilate us, but only insist that we share our duration with them. This is a possibility I explore in this book's coda.

*Chapter 1*

# Composing Human Time

## Locke, Hume, Addison, and Diderot

FINDING IT IMPOSSIBLE to define time through intellectual inquiry, St. Augustine turns to experiencing the reverberations of his own duration in a prayer which he compiles of quotations from various psalms. The move to a self-evident existential duration entails for Augustine not only naming poetry and voice, but also performing both the act of composing and the act of reciting. When duration cannot be defined, it is exemplified in artifacts constituted by compositional activity and sensible experience.[1] In this chapter I trace a similar move in a number of eighteenth-century English discussions of time where the focus on how we come to know duration turns into the suggestion that more than coming to know it, we come to feel it, and that we come to feel it when we listen to music, or read essays and novels, or converse with friends. Music, novels, and conversation—like Augustine's prayer—solicit the senses as well as the capacity to recognize and form temporal patterns.

A while ago Georges Poulet argued that eighteenth-century empiricists came to understand temporal continuity as a human fabrication arising from the ways in which our minds seek to integrate instances of sensory experience. For the empiricists, Poulet explained, "intensity of sensation ensures the instant; multiplicity of sensation ensures duration."[2] Poulet's compact yet wide-ranging study offers an overview of the emergence of temporal phenomenology from the Renaissance through the early twentieth century. He argues that the break with traditional Christian paradigms launched by the Renaissance introduced duration as a conceptual problem that gave rise to a burgeoning discourse—both philosophical and literary—on the nature of time and its relation to experience. Endurance came to be understood as a conceptual difficulty that calls for human—rather than divine—explanations. In Poulet's account, seventeenth-century discussions conceived of such human time as a succession of durationless instants whose continuity must be repeatedly asserted from without. For "the seventeenth-century man," he writes, "duration is a chaplet of instants. The creative activity alone permits passage from one bead to another" (14). By contrast, eighteenth-century discussions began to understand continuity as in and of itself a human fabrication and to examine the possibility of a more organic relation between instants and duration. And while eighteenth-century culture was concerned with positional relations of disjunctive moments, Romantics conceived of each moment as though it encapsulated a linear span; nineteenth-century writers emphasized causal relations among moments; and twentieth-century authors conceived of the moment as a nondeterministic potential. Poulet's survey remains highly suggestive; it also, however, leaves many specificities unexplored. He usefully emphasizes the importance of intensities and multiplicities in eighteenth-century empiricist discussions of time, but these discussions also persistently raise questions about how such intensities and multiplicities combine to support temporal experience. One pressing question they raise is whether the instant counts as temporal or atemporal. Another is how exactly durational multiplicity relates to and differs from intensification. Yet another regards varying models for integrating multiplicities.

In this chapter I track these questions as they are explored in John Locke's and David Hume's philosophies of time and in Joseph Addison's and Denis Diderot's comments on discursive compositions and durational experience. In these early sensationist and associationist works, we find various models for understanding the qualities of the instant and the compositional organization of durational multiplicities. Locke famously defines our primary temporal experience through the succession of our ideas, but we will soon see that while his explicit argument focuses on measurement of lengths and presumes the instant to be atemporal, the examples he presents point up sensations and intensities that turn the instant into a part of duration and sway the discussion from estimations of quantities to assessments of qualities. In Hume's analyses of time, we find a more direct exploration of such durational qualities not only highlighting the durational intensities of the instant, but also qualifying the associative strings that integrate moments into temporal expanses. And in

Addison's and Diderot's comments on literary aesthetics, we can trace links between durational qualities and an active crafting done by discursive compositions. For Addison and Diderot, durational media such as essays and novels come in numerous genres and forms whose effects variously shape readers' temporal experiences.

## Duration as the Succession of Ideas: Locke and Hume on Quantities and Qualities

For most historians of philosophy, Locke's contribution to understanding time seems minimal: popularizing Isaac Newton's distinction between absolute and relative time while focusing on duration's sensible approximations. Nonetheless, in both popular and academic writing in England throughout the eighteenth century, Locke's discussion of "Duration and Its Simple Modes" in *An Essay Concerning Human Understanding* (1689) was a constant touchstone. As we will soon see, Hume relies on Locke's initial definitions, and Addison frames his notes on time as a popularization of Locke's views. And as we will see in subsequent chapters, Edmund Burke founds his durational aesthetics on both direct and indirect polemics with Locke, and Laurence Sterne quotes and misquotes Locke's *Essay* in his experiments with narrative duration. These appreciative revisions recognize tensions between Locke's arguments and his suggestive discussions of examples, pushing the experiential turn that the *Essay* inaugurates beyond the predominantly quantifying approach that Locke takes. For if in his arguments Locke focuses on the stipulated agreements by which we measure time's expanse, his examples highlight sensorial and emotional dimensions by which we feel the duration of its instants.

Locke relies on Newton's distinction between absolute time—a durable medium that flows uniformly and equably—and relative time—our approximate measures of this medium. But he uses a different set of terms. Locke uses the word "duration" to mean real, objective temporality, which flows uniformly, equably, and constantly. He uses the phrases "sense of duration" or "idea of duration" to mean our primary notion of what duration is—our personal experience of temporal flow as it gauges real external duration. Locke uses "time" to mean a common measure of duration—an abstraction that relies on principles of periodicity. And, finally, he uses "time in general" to mean our approximations of objective duration which we can deduce from the idea of periodicity. But most important, Locke's interest in how we come to know the world through our senses prompts him to focus on personal experiences both individual and collective—what he calls "sense of duration" and "time"—rather than on "duration" as such and "time in general." In this sense, "Locke's epistemology of time is the mirror image of time in Newton's natural philosophy," as Philip Turetzky puts it, investigating time as a dimension of subjective experience more than as an objective medium.[3]

Locke's focus on subjectivity is best illustrated at the outset of chapter xiv in Book II of the *Essay*, where he launches the discussion with his version of the *cogito*: I think therefore I endure. "For whilst we are thinking," he writes, "or whilst we receive successively several *Ideas* in our Minds, we know that we do exist; and so we call the Existence, or the Continuation of the Existence of our selves, or any thing else, Commensurate to the succession of any *Ideas* in our Minds, the *Duration* of our selves" (182).[4] Locke continues to argue that "we have no perception of *Duration*, but by considering the train of *Ideas*, that take their turns in our Understandings. When that succession of *Ideas* ceases, our perception of Duration ceases with it" (182). He thus disengages any direct relation between time and motion and insists, instead, on the routing of our sense of duration through transformations that we register cognitively. Though time is an absolute external substance, Locke contends, our experience of it relies on the mediation of our psychological capacities, rather than on direct access to external motion.

By equating duration with a perception of thinking and with personal endurance, Locke positions his inquiry squarely in the realm of human time and emphasizes individual consciousness.[5] And yet by equating personal endurance with the succession of ideas, Locke opens up his inquiry of time to many of the complications that more generally trouble his examination of human psychology. For the "idea" in Locke's philosophy is a concept both absolutely central and notoriously vexed, as it marks a porous borderline between external realities and internal experience. Arguing against conceptions of human cognition that rely on innateness, Locke insists that all of our ideas rely on our sense perceptions. But he also recognizes that our perceptions are not simply mirror impressions of an external world. Our cognition mediates stimulation such that most of our ideas are very different from the properties of objects. To begin with, Locke distinguishes between primary and secondary qualities of objects and argues that while our ideas of the first resemble objects, our ideas of the second arise from our sensory apparatuses as these reshape and remake the stimulation emanating from objects. As he describes it, "The particular *Bulk, Number, Figure, and Motion of the parts of Fire, or Snow, are really in them*, whether any ones Senses perceive them or no: and therefore they may be called *real Qualities*, because they really exist in those Bodies. But *Light, Heat, Whiteness*, or *Coldness, are no more really in them, than Sickness or Pain is in* Manna. Take away the Sensation of them; let not the Eyes see Light, or Colours, nor the Ears hear Sounds; let the Palate not Taste, nor the Nose Smell, and all Colours, Tastes, Odors, and Sounds, as they are such particular *Ideas*, vanish and cease, and are reduced to their Causes, *i.e.* Bulk, Figure, and Motion of Parts" (137–38). Color, taste, smell, and sound, Locke explains, do not exist without eyes, tongues, noses, and ears to perceive them. Secondly, attention determines which stimulations register in our understanding to produce ideas and which do not: "A sufficient impulse there may be on the Organ; but it not reaching the observation of the Mind, there follows no perception: And though the motion, that uses to produce the *Idea* of Sound, be made in the Ear, yet no sound is heard. Want of Sensation in this case, is not through any defect in the Organ, or that the Man's Ears are less affected, than at other times, when he does hear: but that which uses to produce the *Idea*, though conveyed in by the usual Organ, not being taken notice of in the Understanding, and so imprinting no *Idea* on the Mind, there follows no Sensation" (144). If we are already absorbed in a thought of one kind, then any amount of stimulation around us that is irrelevant to that thought will go unnoticed. Finally, memory guarantees that we can have ideas of objects in their absence—that we can have ideas even when our senses do not deliver any external stimulation. As Locke explains, "This *laying up* of our *Ideas* in the Repository of the Memory, signifies no more but this, that the Mind has a Power, in many cases, to revive Perceptions, which it has once had, with this additional Perception annexed to them" (150). And he concludes: "*Memory*, in an intellectual Creature, is necessary in the next degree to Perception. It is of so great moment, that where it is wanting, all the rest of our Faculties are in a great measure useless" (153). Memory turns out to be, in Locke's account, a supplier of ideas no less than external objects, making perceptions of the external world—once it has been perceived—no longer necessary.

Our cognition thus functions as a mediation that variously shapes and makes ideas and, even more important for my purposes here, shapes and makes succession. Attention, for example, crucially determines the speed and content of the succession of our ideas. If our attention is fixed by one idea, this halts succession, which in turn, by Locke's initial definition, slows down our sense of duration or even brings it to a standstill. While objective duration may flow equably and uniformly, our primary sense of duration cannot be as steady or regular. And if memory supports the generating of ideas, then this suggests that our succession of ideas may not only, or not simply, be moving forward. If our sense of duration is determined by the succession of ideas, then an idea retrieved from memory supports the experience of time passing from one moment to the next by thrusting us back into a moment that has already passed. Thus even as objective time can only flow from past to future, our primary sense of duration cannot be unidirectional. Moreover, Locke's account of the mediating role of our cognition entails that any enhancement or diminishment of our capacities—an extraordinarily sharp memory or, conversely, memory loss; attention deficit or, conversely, hyperfocus; the use of technological tools such as a microscope or, conversely, sensory disabilities such as blindness—could radically challenge any generalization a philosopher might make about ideas and their succession.[6] For Locke's inquiry of time, such complications raise two questions. First, how might shared measures and collective experiences of duration arise from the indeterminate and widely varying conditions that generate the succession of ideas in individual minds? And second, what kind of a sense of endurance might we have when our attention, senses, or memories are disabled or significantly altered?

The first question is especially pressing, for throughout his discussion Locke emphasizes our need for shared public measures of time. The mind, he argues, naturally searches for "some *measure of* this common *Duration*, whereby it might judge of its different lengths, and consider the distinct Order, wherein several things exist, without which a great part of our Knowledge would be confused, and a great part of History be rendered very useless. This Consideration of Duration, as set out by certain Periods, and marked by certain Measures or *Epochs*, is that, I think, which most properly we call *Time*" (187). But how do we arrive at such "Measures or *Epochs*"? How might we agree on such standards if our primary sense of duration varies contingently? Just as Locke does not explain how we construct and arrive at consensus about language or money, he also does not offer detailed conjectures on how we construct and arrive at consensus about temporal measures. But he does suggest that there are, after all, ways for us to asses durations we do not experience, and this makes possible agreements about conventional measurement. If we need not consult our widely varying experience in our assessments of time, then consensus about extensities becomes much more readily available.

There are many occasions in which we do not experience duration as it passes, Locke points out, the most common of which is sleep. And so he explains: "Though a Man has no Perception of the length of Duration, which past whilst he slept or thought not: yet having observed the Revolution of Days and Nights, and found the length of their Duration to be in Appearance regular and constant, he can, upon the supposition, that that Revolution has proceeded after the same manner, whilst he was asleep or thought not, as it used to do at other times, he can, I say, imagine and make allowance for the length of *Duration*, whilst he slept" (183). We may remember that night follows day, follows night again, but in order to recognize "the length of their Duration to be in Appearance regular and constant"—which, of course, contrary to Locke's assertion, they are not—we must have already abstracted a sense of measure from their occurrence. This abstracted idea of periodicity serves a double function: first to validate the observation that at stake is not simply succession but also, more specifically, equable succession, and second to deduce how long we have slept. Locke readily acknowledges how the abstracted idea of periodicity serves the second function—how we "imagine and make allowance for the length of *Duration*, whilst he slept." The first function of our abstracted idea of periodicity—that we "found the length of their [Revolution's] Duration to be in Appearance regular and constant"—remains only indirectly implied. But once we recognize that we would need an abstract measure to assess the revolutions of day and night as recurring at the same intervals, we must also acknowledge that the move from our "sense of duration" to "time"—that common measure in Locke's vocabulary—involves the superfluity of our primary sense of duration: in order to measure the length of succession we need a standard independent of our experience of this succession. It thus turns out that counting time relies on abstractions that may or may not validate experiences but that do not arise from these experiences. Measures of time, then, rely on stipulated standards that make an actual primary experience of duration irrelevant.

Throughout Locke's discussion we find such tension between a reliance on abstractions for measuring time and a lingering commitment to sensation. On the one hand, Locke points out the arbitrariness of our conventions. "*Minutes, Hours, Days, and Years*," he reminds us, "are then *no* more *necessary to Time* or Duration, than Inches, Feet, Yards, and Miles, marked out in any

Matter, are to Extension" (191). On the other hand, he privileges those conventions that seem as though they are verified by direct sensory evidence. Thus, for example, he prefers the temporal estimations of a blind man who can smell, taste, and feel seasonal change over those generated by calendars:

> Thus we see that Men born blind, count Time well enough by Years, whose Revolutions yet they cannot distinguish by Motions, that they perceive not: And I ask, whether a blind Man, who distinguished his Years, either by heat of Summer, or cold of Winter; by the Smell of any Flower of the Spring, or taste of any Fruit of the Autumn, would not have a better measure of Time, than the *Romans* had before the Reformation of their *Calendar* by *Julius Cæsar*, or many other People, whose Years, notwithstanding the motion of the Sun, which they pretend to make use of, are very irregular: and it adds no small difficulty to Chronology, that the exact lengths of the Years that several Nations counted by, are hard to be known, they differing very much one from another, and, I think, I may say all of them, from the precise motion of the Sun. (189)

And yet, the heat of air, the blossoming of flowers, and the ripeness of fruit are no less "irregular" than "the motion of the Sun" that underwrites the old Roman calendar, which Locke here criticizes. What drives Locke's privileging of the blind man's temporal estimations in this passage is the sheer sensual abundance of his description. That the blind man relies on three senses that connect his estimations of the passage of time to nature's seasons suggests to Locke that he can "count Time well enough"—even better than those whose estimations of the sun's motions in their calendars turn out to be faulty. The fewer mediating apparatuses between our perceptions of the external world and our temporal estimations of length, Locke indicates, the more reliable our estimations of time will be.

That Locke prefers notions of time passing that more directly arise from sensation becomes even clearer as he proceeds to discuss the measures afforded by the new technology of pendulum clocks.[7] He writes:

> Though Men have of late made use of a Pendulum, as a more steady and regular Motion, than that of the Sun or (to speak more truly) of the Earth; yet if any one should be asked how he certainly knows,

> that the two successive swings of a Pendulum are equal, it would be very hard to satisfie himself, that they are infallibly so: since we cannot be sure, that the Cause of that Motion which is unknown to us, shall always operate equally; and we are sure, that the Medium in which the Pendulum moves, is not constantly the same: either of which varying, may alter the Equality of such Periods, and thereby destroy the certainty and exactness of the measure by Motion, as well as any other Periods of other Appearances, the Notion of Duration still remaining clear, though our measures of it cannot any of them be demonstrated to be exact. (190)

It is especially striking that Locke should emphasize the unreliability of a pendulum in ordinary conditions of experience immediately after privileging the blind man's sensual estimations of time passing. The majority of the passage points out that in most circumstances a pendulum will not deliver to our senses an idea of regular motion. For this reason, Locke concludes, temporal measures that rely on a pendulum cannot "be demonstrated to be exact." And yet, he continues to insist, "the Notion of Duration still remain[s] clear"—as though our sensual experiences of day and night, heat, and ripening fruit, have already settled everything we might want to know about the passage of time. The tracking of seasonal change mediated only by the personal senses seems to him more reliable than any knowledge mediated by technologies whose workings many of us cannot directly observe or understand. This, Locke indicates, makes measures not only imprecise but also experientially suspicious. What Locke less than fully acknowledges here, but that arises indirectly and yet persistently from his discussion, is that the experiential confidence that we have about duration might be best described as feelings of intensities—of temperature, of fragrance, of sweetness.

I am suggesting that Locke's privileging of firsthand sensual experience in his discussions of the various ways in which our succession of ideas supports conventions for assessing temporal length indicates a dimension of time that his concern for and vocabulary of counting eludes. If sensual experience lends certainty to our sense of duration where abstract measures cannot, then the certainty it furnishes is one that assesses qualities more than quantities—one that gauges feelings and intensities rather than the counting of expanses. Such qualitative dimensions of our sense of duration become even more apparent when we closely examine Locke's discussion of durations that our sensory apparatuses are not calibrated to perceive. He points out that a succession of moving objects may happen at speeds that are either too fast or too slow for us to notice. The express purpose of the section is to argue that our primary sense of duration arises from the internal succession of ideas, rather than from our registering of external moving objects. But as Locke presents an example, he turns to redefining the instant in such a way that incorporates it as a part of time even as its extensity cannot be measured.

Here is what Locke has to say about the situation of being shot—his example of succession too fast for us to register cognitively:

> Let a Cannon-Bullet pass through a Room, and in its way take with it any Limb, or fleshy Parts of a Man; 'tis as clear as any Demonstration can be, that it must strike successively the two sides of the Room: 'Tis also evident, that it must touch one part of the Flesh first, and another after; and so in Succession: And yet I believe, no Body, who ever felt the pain of such a shot, or heard the blow against the two distant Walls, could perceive any Succession, either in the pain, or sound of so swift a stroke. Such a part of Duration as this, wherein we perceive no Succession, is that which we may call an *Instant*; and is *that which takes up the time of only one Idea* in our Minds, without the Succession of another, wherein therefore we perceive no Succession at all. (185)

Locke's assertion that "no Body, who ever felt the pain of such a shot, or heard the blow against the two distant Walls, could perceive any Succession, either in the pain, or sound of so swift a stroke" serves a crucial double purpose here. On the one hand, Locke argues that without the perception of succession such an instance can only yield an instant—"*that which takes up the time of only one Idea* in our Minds." On the other hand, his prose highlights a succession of dramatic events and their strong impact on the body—the "pain" and the "blow"—which cannot but evoke some temporal dimension through which and in which the wounded person experiences his endurance. And it is precisely this drama—the event of being shot, highlighted through discursive succession even as, or precisely because, the represented body cannot gauge this succession—that leads Locke to redefine the instant as "a part of time." Recall that the opening paragraphs of his discussion of time define our primary sense of duration as succession—as the difference between one idea and another:

"We have no perception of *Duration*, but by considering the train of *Ideas*, that take their turns in our Understandings. When that succession of *Ideas* ceases, our perception of Duration ceases with it" (182). By this initial logic, the instant, the single idea, is atemporal. However, by the second definition—the instant in the example of being shot—a single idea is already a temporal element—"*that which takes up the time of only one Idea*." In the next chapter Locke reiterates the second definition, the one that conceives of the instant—which here he calls a moment—as a part of duration. "Such a small part in Duration," he writes, "may be called a *Moment*, and is the time of one *Idea* in our Minds" (203). Thus the example of being shot prompts Locke to revise his initial account of our primary sense of duration to include the instant. And this revision suggests, in turn, that we might have an idea of duration that does not pertain to extensity and cannot be measured. Though duration can't be counted when we are shot or when we hold a single idea in our minds, though we cannot tell time's length from firsthand sensations on such occasions, we do have some temporal notion—the instant or the moment—crucial to our persisting endurance through such events. And if the quantity or extensity of such moments cannot be assessed, their quality or intensity seems unavoidable. At stake here is an immeasurable moment that cannot be subtracted from endurance. I think therefore I endure—"the Continuation of the Existence of our selves" with which Locke began his chapters on duration—thus turns out to be "Commensurate to the succession of any *Ideas* in our Minds," with an emphasis falling on "any" more than on "succession." Any single idea turns out to be key to our sense of duration insofar as its qualities inescapably graft themselves on our minds and bodies and even as its quantities remain unknown.

Locke's discussion, then, points up qualitative dimensions of durational experience that his explicit argument cannot account for. These qualitative dimensions arise from irregularities of succession and from intensities of sensations, aspects of "the succession of ideas" whose impact on our sense of duration Locke considers only indirectly. In Hume's *A Treatise of Human Nature* (1738), however, these dimensions take center stage. Far more than Locke, Hume focuses on differentiating between "degrees and force of liveliness," as Hume puts it throughout the first chapter of the *Treatise*. Hume distinguishes between impressions and ideas and devotes much attention to the passions, which he takes to deliver impressions of the strongest degree, in contrast to "the faint images of these in thinking and reasoning," (1), which he calls "ideas."[8] Hume also examines the logic of association among impressions and ideas much more thoroughly than Locke does. Whereas Locke posits succession as a singular principle requiring little analysis and devotes only one chapter to the association of ideas in which he primarily focuses on the dangers of chance connections, Hume argues that resemblance, contiguity, and cause and effect are all principles that universally and naturally encourage our relating one idea or impression to another thus complexly shaping thoughts and feelings (11).[9] Hume proposes that such associative operations are responsible for both the diachronic movement of our minds—the succession of our ideas and impressions—and the synchronic resonances that reverberate in the mind at a single instant—the complex ideas and passions that yoke together a number of impressions and ideas into coherent notions of substance or mode.[10] Furthermore, he distinguishes between the ways by which ideas connect to one another and the ways by which impressions do, arguing that "ideas are associated by resemblance, contiguity, and causation; and impressions only by resemblance" (283). For Hume, then, our minds are constituted to begin with by a variety of compositional principles, thus yielding varying qualities of association. Together these principles of association form "a kind of Attraction, which in the mental world will be found to have as extraordinary effects as in the natural, and to shew itself in as many and as various forms" (12–13).

Such privileging of sensible, affective, and compositional dimensions informs Hume's discussion of time in Book I of the *Treatise*. Hume argues "that time cannot make its appearance to the mind, either alone, or attended with a steady unchangeable object, but is always discover'd by some *perceivable* succession of changeable objects" (35). But Hume directly recognizes that attention to such succession of perceptions yields notions not only of quantity, but also of quality. Hume defines "the idea of time" (34) as "an abstract idea, which comprehends a still greater variety than that of space, and yet is represented in the fancy by some particular individual idea of a determinate quantity and quality" (35). Qualities of duration arise because, according to Hume, our idea of time is "deriv'd from the succession of our perceptions of every kind, ideas as well as impressions, and impressions of reflection as well as of sensation" (34–35), thus including "internal impressions," which he specifies as "our passions, emotions, desires and aversions" (33) and which he has earlier excluded from our idea of space. Moreover, Hume emphasizes that each and every one of the impressions and ideas that combine into our sense of time constitutes a substantial experience: " 'Tis certain then, that time, as it exists, must be compos'd of indivisible moments" (31) that "must be fill'd with some real object or existence" (39), which he also calls "sensible qualities" (39). He does not count these "indivisible moments" as supporting our sense of time; following Locke's explicit argument, only succession does this job for Hume. Yet he devotes much of his discussion to the experiential endurance of these units, highlighting their positivity. Hume defines the moment in Book I of the *Treatise* as, strictly speaking, not a part of time; and yet, he also attributes to it what Gilles Deleuze identifies in Hume's philosophy as "real existence." As Deleuze explains it, "real existence" is "neither a mathematical nor a physical point, but rather a sensible one," and he adds, "a physical point is already extended and divisible; a mathematical point is nothing." A sensible point is indivisible, and yet it is something.[11]

Beyond highlighting the moment as a part of endurance by virtue of its "sensible qualities," Hume argues that our idea of time is not an object in and of itself, "but arises altogether from *the* manner, in which impressions appear to the mind" (36). A moment "must be fill'd with some real object or existence," and our idea of time is created by the very principles of composition that string these real objects together. And if these principles of composition are by Locke's analysis mere "succession," then Hume's analysis differentiates variable forms of succession. More specifically, Hume exemplifies succession as the order and rhythm in which musical notes appear, suggesting that even as principles of succession are not objects in and of themselves, they cannot be extricated from the existences through which they are perceived. Thus, Hume explains, "Five notes play'd on a flute give us the impression and idea of time; tho' time be not a sixth impression, which presents itself to the hearing or any other of the senses. . . . But here it only takes notice of the *manner*, in which the different sounds make their appearance" (36–37). Such an explication of the succession that defines time produces it as an aesthetic form—both a logic of composition and an ineluctably experiential apprehension of it.

When Hume discusses measurement, he considers it to be more like an artist's intuition inextricable from his practice than like a mathematical abstraction. In a long discussion of the unreliable precision of all our measures of size—of bodies and space as much of as of time—Hume comments about the aptness of our intuitive estimations of duration, analogizing these to artists' intuitions about their media. He writes:

> This appears very conspicuously with regard to time; where tho' 'tis evident we have no exact method of determining the proportions of

> parts, not even so exact as in extension, yet the various corrections of our measures, and their different degrees of exactness, have given us an obscure and implicit notion of a perfect and entire equality. The case is the same in many other subjects. A musician finding his ear become every day more delicate, and correcting himself by reflection and attention, proceeds with the same act of the mind, even when the subject fails him, and entertains a notion of a compleat *tierce* or *octave*, without being able to tell whence he derives his standard. A painter forms the same fiction with regard to colours. A mechanic with regard to motion. To the one *light* and *shade*; to the other *swift* and *slow* are imagin'd to be capable of an exact comparison and equality beyond the judgments of the senses. (48–49)

Hume argues that we have strong intuitions about temporal quantities that rely on repeated practice, which belie both theoretical understanding and the contingencies of sensory experience. Our estimations of temporal measures, he suggests, are irreducibly practical, arising from doing more than from knowing, from practice more than theory. But if, say, a violinist's practice is constituted by repeated performances of a number of musical pieces, what counts as an ordinary time-teller's practice? Hume begins the paragraph indicating that consulting our watches might be the practice at stake: "The various corrections of our measures, and their different degrees of exactness, have given us an obscure and implicit notion of a perfect and entire equality," he writes. And as Stuart Sherman has documented meticulously, the technology for individual watches and clocks, which was developed in the seventeenth century, became widely available for middle class consumers in the eighteenth century, turning chronometric measuring into a general practice. But if we allow Hume's analogies the power of illuminating the principle that they compare with, then we might notice that in these Hume does not highlight steady beat or determinate lines or parts—elements of musicians', painters', and mechanics' craftsmanship that more obviously support a numerical estimation of proportion and are thus more obviously analogous to watches and clocks. What Hume is after in invoking a musician's ear for an octave, a painter's eye for color, and a mechanic's sense of motion are discriminations that rely on contrasts rather than on more minutely expressed comparisons but that nonetheless deliver fine-grained assessments. While the sources of these qualities are precise mathematical proportions, the artist gauges them not as such, but rather as the sum of a sensual effect—a particular shade of turquoise rather than blue, or a C-major scale rather than a D-major scale. Thus Hume directs us to think of our temporal assessments more along the lines of estimations of qualities than approximations of chronometry—school time feels different from commute time which feels different from dinner time. Even when Hume discusses temporal measure and extensity, then, he describes these as intuitions comparable to feelings about intensities of sensual effects more than to estimations of length.

Hume returns to questions of time in his examination of the passions in Book II of the *Treatise*, where he devotes chapters vii and viii of Part III to examining how spatiotemporal position variously shapes our feelings towards objects. If the discussion in Book I stresses that duration has various qualities that arise from varying arrangements of impressions and ideas (whether simultaneous or in succession), then here the varying qualities of experience are determined by the spatial and temporal arrangement of objects in relation to us. Hume argues, for example, that because we can perceive distant space coextensively with ourselves but the experience of time is necessarily successive, it is harder to think vivaciously of distant times than of distant places. Additionally, because the past flows away from us while the future gets closer to us, it is easier to imagine a future than a past. Such observations may seem surprising for twenty-first-century readers who tend to privilege the ways in which the succession of our ideas repeatedly connects up with memories—whether as compulsive and disturbing trauma, or as sweeter and more deliberate nostalgia. But for Hume—and as we will see in the next chapter, for Richardson and Hutcheson as well—a succession of ideas that aims to project the future seems like an especially valuable ethico-cognitive activity. But in any case, even as we may or may not agree with the content of Hume's observations about our feelings toward past and future, we should recognize the extent to which he insistently attends to qualities and to the way by which such qualities are generated by compositional determinations. When examining varying strengths of feeling, Hume considers their organization in time; when defining time, Hume thinks about aesthetic qualities that arise from organizational patterns. Throughout the *Treatise* we find a homology between sensibility and temporal process, and between particular varying manifestations of sensibility and particular varying patterns of temporal process.[12]

## Crafting the Succession of Ideas: Addison and Diderot on the Pleasures of Reading

In a beautiful examination of the way walks—actual physical movement along varying architectural paths—shape Addison's model of the mind, Sean Silver discovers recurrent interchanges of design and nature, representations and originals. While the walk as a whole follows a predesigned path, certain moments of prospect invite the walker to digress on his own reflective rambles. Good writing in Addison's estimation works similarly, according to Silver, with a well-ordered composition that includes well-placed opportunities for readers to digress on their own imaginative excursions. "These 'hints' are the opportunity, as the etymology of the words suggests, for the reader to be 'seized' by a moment; they activate the imagination's secondary pleasures, those pleasures that occur when 'the imagination takes [a] hint'—or, we might suspect, is *taken* by one—'and leads us unexpectedly into cities or theatres, plains or meadows.'"[13] This double principle of the workings of the mind, Silver argues, aligns Addison with his contemporary associationists. In Addison's model the task of the poet is "to create an image in such a way that it prompts a rich chain of pleasurable associations in the mind of his anticipated reader" (140). Addison's "empiricist epistemological model adapted to an aesthetics" (136), then, conceives of the mind's labors through the two principles of well-organized encasement and materialist digression that Silver also finds in Hooke, Locke, and many others.

For my purposes here, Silver's appraisal of Addison is especially valuable, as it brings us very close to recognizing the temporal dimensions of Addison's aesthetics. For in the tradition of associationist empiricism, entrusting the poet with eliciting a succession of ideas also entrusts him with mediating readers' durational experience. But, more like Hume than like Locke, Addison explicitly approaches duration as variously patterned and intensely qualified. And he emphasizes how we can fashion it as such, conceiving of qualitative durational experience as a consequence of crafting. For Addison human time, while contained within an equable flow, does not at all resemble it and is actively and variously shaped by reading materials that prompt varying associative chains.

In his *Spectator* 476 (5 September 1712), Addison identifies two styles of writing, each differently shaping the experience of its readers by way of the form of succession it promotes:

> When I read an Author of Genius, who writes without Method, I fancy myself in a Wood that abounds with a great many noble Objects, rising among one another in the greatest Confusion and Disorder. When I read a Methodical Discourse, I am in a regular Plantation, and can place my self in its several Centers, so as to take a view of all the Lines and Walks that are struck from them. You may ramble in the one a whole Day together, and every Moment discover something or other that is new to you, but when you have done you will have but a confused imperfect Notion of the Place; in the other, your Eye commands the whole Prospect, and gives you such an Idea of it, as is not easily worn out of the Memory. (4:186)[14]

The nonmethodical discourse gives pleasures that are physical and serial; it is analogized to a day hike in which visibility is limited but surprising new discoveries and motion abound. The methodical discourse gives pleasures that seem at first to be more purely cognitive and synchronic; it is analogized to the static observation of architecture. The nonmethodical discourse shares a temporal medium with the vehicle of its simile, thus emphasizing duration, while the methodical seems initially to promote the illusion of atemporality by virtue of being coupled in a simile of a spatial, rather than a temporal, medium. And yet Addison begins to suggest the temporal stakes of even the methodical discourse by casting its value in terms of the traces it leaves for future recall. If "your Eye commands the whole Prospect," this matters only because it "gives you such an Idea of it, as is not easily worn out of the Memory." This suggests that while the nonmethodical discourse supports the succession of ideas in something like a continuous present—"a whole Day together"—the methodical essay supports the succession of ideas in the future; as the future becomes present, one can connect one's present ideas to past ones more readily with a methodical essay in store.

Addison continues to develop the value of the methodical discourse by further elaborating on the temporalized trajectories of thought it supports. At stake is not only future recall, but also successive strings issued at present: "When a Man has plann'd his Discourse, he finds a great many Thoughts rising out of every Head, that do not offer themselves upon the general Survey of a Subject. His Thoughts are at the same time more intelligible, and better discover their Drift and Meaning, when they are placed in their proper Lights, and follow one another in a regular Series, than when they are thrown together without Order and Connexion" (4:186). The methodical composition invites "a great many Thoughts rising out of every Head"—successions of ideas that "do not offer themselves upon the general Survey of a Subject" and thus count as digressive, but that are "at the same time" reigned in to discover the "Drift and Meaning" of the author, proceeding in the direction of a future projection for the argument at hand.

Thus both the nonmethodical and methodical compositions promote a pleasure to be had by eliciting and shaping the succession of ideas of their readers, but each uses a different pattern that, in turn, supports different emotions. The first—the rambling walk of an essay—is composed as a meandering line and elicits the feeling of pleasing confusion; the second—organized as a set of points of view—is composed more like a spiraling line that loops forward and back as it also proceeds steadily and gives a reassuring feeling of command. In Chapter 4 we'll see how these different forms of lines are adapted in later discourses of durational aesthetics—the first develops into the haptic beauty of Hogarth's serpentine lines and Burke's curved bodies and is then adapted into the ultimately safe pleasures of Ann Radcliffe's gothic lines, while the second anticipates Radcliffe's crucial flagging of recollection in her accounts of how a mind achieves pleasing composure. For now it suffices to recognize the extent to which temporality permeates Addison's understanding of the pleasure of reading. Whether likened to a rambling walk in the woods or to an ocular prospect of a plantation, reading supports the succession of ideas of readers and gives them a sense of duration. They may "every Moment discover something or other that is new," or they may have "a Drift" that carries "their own meaning" that also anticipates the meaning of the author. But both models of composition, Addison insists, occasion a succession of ideas and as such shape their readers' experience of duration into variously formed and felt qualitative experiences.

But if I have been arguing that Addison entrusts leisure reading with the task of mediating qualitative time, then on many accounts reading in the eighteenth century has done nothing but recruit leisure for the purposes of a rational culture with its economic priorities and its chronometric consciousness. Middle-class life in eighteenth-century England had come increasingly to lay stress on a differentiation between work and leisure, which, as Patricia Meyer Spacks points out, "implies a principle of differentiating time."[15] J. H. Plumb documents how the print industry and literacy helped to differentiate work from leisure. He argues that the explosion in production and dissemination of print was especially remarkable because reading materials could be picked up almost anywhere and anytime, and because it aided the commercial organization of other industries. But Plumb also argues that the difference between work and leisure was quickly eroded by the expansion of work's utilitarian morality and chronometric rhythms into practices of leisure. He delineates how leisure activities came to be conceived as an exertion to increase social standing, or what we might call cultural capital. Similarly, Franco Moretti explains that leisure came to be moralized as work, and such an approach privileged recursive and systematic forms of writing that rationalize fictional representations, thus making them analogous to an economized public sphere. And Deidre Lynch concludes that one result of such reconceived utility of leisure is that "literature begins to be reconceptualized as a steadying influence on those who love it."[16] These studies highlight the reliable periodicity of entertainment production and especially of print publication as the most important dimension of the durational experience that leisure activities promote. From the perspective of such periodicity, entertainment looks like a version of clockwork, extending the chronometric logic of usefulness and work into the durational experience of leisure.

Addison, to be sure, endorses the chronometric approach throughout his writings. In *Spectator* 10 (12 March 1711), he famously addresses "all regulated Families, that set apart an Hour in every Morning for Tea and Bread and Butter; and would earnestly advise them for their Good to order this Paper to be punctually served up, and to be looked upon as a Part of the Tea Equipage" (1:44–45). Sherman persuasively details Addison's contribution to the fashioning of a chronometric consciousness based on such recurring invocations of the reliable periodicity of journal publication. In these contributions, Sherman explains, Addison promotes "a diurnal paradigm for achieving, recognizing, and inhabiting the fullness of time."[17] But while Addison routinely makes the case for the potential usefulness of reading, he also often insists that both this usefulness and this time rely on a different logic from the economized values of work. In *Spectator* 287 (29 January 1712), he emphasizes that entertainment responds to needs other than our ambitions in work: "We look out for Pleasures and Amusements; and among a great number of idle People, there will be many whose Pleasures will lie in Reading and Contemplation" (3:21), he writes. And he often considers the pleasure of reading and contemplation as a pause from the demands for action constantly pressing on us in daily life; aesthetic pleasure promotes a special "time or leisure to reflect" (3:569), he argues in

*Spectator* 418 (30 June 1712). Such discussions not only chart real differences between various kinds of ordinary activities but also differentiate among genres of leisure reading by the varying ways form directs the succession of ideas.

Addison makes this point most directly in his notes on time in *Spectator* 94 (18 June 1711), where he takes up Locke's primary definition of duration from the *Essay*, while also suggesting that Locke's analysis begs for a distinction between quantities and qualities. If, as Michael Ketcham argues and as subsequent scholars have tended to accept, "*The Spectator* in effect dramatizes Locke's account of duration," then this dramatization serves to think through those details of quality whose tension with Locke's focus on measures relegates to the status of implications rather than propositions.[18] Addison begins by defining duration as the succession of ideas and proposing to "carry this Thought further, and consider a Man as, on one Side, shortening his Time by thinking on nothing, or but a few things; so, on the other, as lengthening it, by employing his Thoughts on many Subjects, or by entertaining a quick and constant Succession of Ideas" (1:399). But soon Addison points out the improbability of our cogitations extending our lives, and he shifts the terms of the discussion from measuring the expanse of any given duration to assessing the feelings it generates. He presents an anecdote about a sultan who immerses his head in water for a few seconds and experiences an alternate life in which he achieves as much as rising, falling, marrying, and giving birth to seven sons and daughters. Such "Eastern Fables" (1:401), Addison explains, illustrate a cosmological truth: God's eternal nature can scramble human time at will, "mak[ing] a single Day, nay a single Moment, appear to any of his Creatures as a thousand Years" (1:401). But to Mr. Spectator and his English readers, contemplating alternate durations yields different conclusions—conclusions more adequate to their own experiences of the empirical world: "The Hours of a wise Man are lengthened by his Ideas, as those of a Fool are by his Passions: The Time of the one is long, because he does not know what to do with it; so is that of the other, because he distinguishes every Moment of it with some useful or amusing Thought; or in other Words, because the one is always wishing it away, and the other always enjoying it" (1:401). Thus the lesson of *Spectator* 94 is to keep oneself busy with leisure reading such as that offered by Addison's journal, but not because a reader lives longer than the fool or even because he *feels* as though he does. Both endure and experience their durations as having a similar expanse, but the wise man enjoys his time while the fool is bored with it.

One purpose of the eastern fable within this number is to contrast a probabilistic chronometric assessment of time—how long actions must take compared with the length of time allotted to them in a narrative—with a qualitative assessment—the feelings a narrative generates. One moment with one's head under water may feel like a hundred years; this defies a reader's probabilistic expectations, but it does generate awe-inspiring wonder. Yet as we have seen, Addison also insists that the chronometric is far from the only temporal logic to shape his writing and to affect his readers' durational experience. And thus the eastern fable of *Spectator 94* also serves a second purpose: to suggest how different aesthetic forms varyingly construct durational qualities for their readers. The eastern fable transports readers to other times and places—from England to the East, from witnessing a head momentarily immersed in water to following an entire lifetime. *Spectator* essays, alternatively, transform readers' own time—from dull fixations or passionate eruptions to more safely enjoyable pleasures, or, as Addison puts it in the beginning of the essay, "Thoughts on many Subjects . . . entertaining a quick and constant Succession of Ideas." And with this Addison indicates that duration varies not only by whether we approach it quantitatively or qualitatively, but also by how different compositional principles shape our experience of it. The genre of the eastern fable mediates for readers other-worldly transport replete with awe-inspiring achievements; the genre of the periodical essay mediates for readers more ordinary emotional qualities.

Addison's discussion is, of course, Orientalist in Edward Said's sense, using an "eastern fable" as a vehicle for formulating insights about Addison's own English sense of time. As Srinivas Aravamudan points out, the oriental tale was extremely popular in England in the early and mid-eighteenth century, "trolling for the same readers" as realist genres, which suggests that Addison uses this anecdote to comment about English readers' expectations and experiences of their leisure reading materials more than about those of easterners.[19] Moreover, scholars have recently presented the contrast that Addison maps geographically as a historical succession of chronotopes—from the emblematic to the chronometric, and from the romance to the novel. The temporal conception of Addison's "East" resembles in crucial ways the Christian framework of eternity and its emblematic resonances of moments, even as it is being supplanted by a historicist paradigm organized by linear succession in rational grids.[20] It also resembles romance's spatiotemporal transports that leave no trace on psychologies and bodies, to recall Mikhail Bakhtin's analysis of adventure time and Patricia Parker's discussion of digressions, which I discuss in the Introduction. But in Addison we can already begin to see how sensibility refigures romance, as the spatiotemporal transports of fable are undermined by the emotional transports of the periodical essay, which may be equally digressive but, nonetheless, retain a strong connection to people's ordinary experiences in their ordinary lives.[21]

As I argue in the Introduction, the sensibility chronotope adapts, rather than more simply negates and supplants, the romance chronotope. We can see especially well such adaptation at work in Diderot's essay "In Praise of Richardson" (1762). Diderot combines Addison's genres of the eastern fable and the periodical essay by describing a probable novel in the vocabulary of romance abundance, while transforming spatiotemporal transports into purely durational intensities—diverse qualitative temporal experience. When reading Richardson's novels, Diderot explains, "In the space of a few hours I had been through a host of situations which the longest life can scarcely provide in its whole course. I had heard the genuine language of the passions; I had seen the secret springs of self-interest and self-love operating in a hundred different ways; I had become privy to a multitude of incidents and I felt I had gained in experience" (83).[22] If a few hours with a Richardson novel offer more intense and varied experience than an average lifetime, then the duration of reading is of such a unique quality that it cannot possibly be assessed by chronometric measures. And yet if such overabundance recalls Addison's eastern fable, then Diderot emphasizes that reading Richardson doesn't take us elsewhere, but rather enables us to experience our own time more intensely: "The world we live in is his scene of action. . . . The passions he portrays are those I feel within me. . . . He shows me the general course of life as I experience it" (83). Addison primarily describes durational experience in an intellectual vocabulary of the succession of ideas—the "useful or amusing Thought" that leisure reading prompts in the wise man of number 94, the "notions," "ideas," "thoughts," and "meaning" that a methodical essay promotes in number 476, and, more generally, "the time and leisure to reflect" of number 418. Diderot overhauls the vocabulary of thinking into the vocabulary of feeling, as might be expected of a sentimentalist of the midcentury for whom, as Jessica Riskin reminds us, "Sensibility was the 'first germ of thought' and 'the most beautiful, the most singular phenomenon of nature.' "[23] But even as for Diderot thinking and feeling are continuous, it is precisely the shift from one to the other that emphasizes the extent to which the incorporation of romance into a specifically durational aesthetics is at stake. For when the intensities of adventure (a key to romance) are rewritten into the intensities of feeling in novels such as Richardson's, then we can confidently inhabit a sensibility chronotope that does not violate spatiotemporal probabilities by insisting that transport belongs with the varieties of qualitative duration. As Diderot follows the characters and plots in Richardson's novels, he barely moves in space or time; but within this more probable stationary position he is moved to feel strongly and, most important, diversely—a succession of feelings that enable us to fully recognize the reappraisal of quantities as qualities and the refiguration of romance into the sensibility chronotope.

In his adaptation of fable to cohere with essay and of romance to cohere with novel, Diderot flags for us the way in which the chronometric and sensibility chronotopes complement one another. But this is far from being his last word about the relation between these different temporal conceptions. For in Diderot's estimation—as in many more recent accounts of the rationalizations of modernity—the quantitative and the qualitative jostle over the same turf. "Take care not to open one of these enchanting works if you have any duties to perform" (85), Diderot begins; and he concludes with pathos: "Emily, Charlotte, Pamela, dear Miss Howe, as I talk with you, the years of toil and the harvest of laurels fall away, and I go forward to my last hour, abandoning every project which might also recommend me to future ages" (97). One might either assess one's exertions by counting the hours, days, years it takes to accomplish a task, or one might immerse oneself in the duration and assess its value by the various feelings it conveys and, thus, lose count of time. But Diderot presents himself as among the few who privilege their qualitative durational experience over the demands of a chronometric public sphere. What Addison identifies as an individual philosopher's oversight—Locke's failure to directly specify qualitative dimensions of time—Diderot suggests is a more pervasive problem of modern culture. Richardson's greatness, he argues, rests not simply on his novels' excellent mimesis of the ordinary but, rather, in their defamiliarization of the ordinary: his novels draw attention to mundane actions' durations to which his contemporaries have been desensitized by chronometric habits. "You accuse Richardson of being long-drawn-out!" Diderot writes, and then returns the accusation: "You must have forgotten how much trouble, care and activity it takes to accomplish the smallest undertaking, to see a lawsuit through, to arrange a marriage, to bring about a reconciliation. Think what you will of these details. . . . They are trivial, you say; it's something we see every day! You are wrong; it is something which takes place every day before your eyes but which you never see" (86). Richardson's novels restore people's sense of the duration of ordinary actions such as marriage and contract, which we no longer seem to notice, and such desensitization arises from overlooking the varying qualities of attention solicited by the durations that these actions inhabit. If a courtship, lawsuit, or reconciliation may be chronometrically summarized to have occupied twelve months, two years, or a decade, such summaries eclipse the abundance of "trouble, care and activity" that one must have experienced while in the midst of these endeavors. With this Diderot flags another way by which sensibility adapts romance: it turns the early genre's loosely stitched digressions into a length necessary for a truer verisimilitude of action as it constitutes character. And finally, Diderot flags the way in which quantitative assessments overrun all attempts to make the qualitative cohere with the quantitative. The quantitative way of knowing, he suggests, habitualizes time so thoroughly that many cannot tolerate Richardson's novels' demands for reactivation of the sensations of original durations.[24]

In the next paragraph Diderot attributes the temporal compression to which his contemporaries have been habituated to market-driven theatrical aesthetics that shorten and thin out audiences' capacities for absorption. "For a people open to a thousand distractions, whose days have not enough hours for all the amusements with which they are wont to fill them, Richardson's books must seem long. . . . You hardly ever go to see more than the last act of a tragedy. Skip to the last twenty pages of *Clarissa*" (86), he remarks sarcastically. An incredibly productive entertainment industry, rather than oversaturating the market, successfully adapts consumers to its needs by limiting their capacities for absorption. Here we have a synchronic jostling between a modern incarnation of the romance chronotope and its refiguration in sensibility. The entertainment industry—with its "thousand distractions" deliberately connoting romance and fabled incredulity—has turned consumers' lives into loosely connected episodes of leisure, making the lengths of their endurance nothing but a series of distractions. Richardson's novels, by contrast, restore us to an inabstractable duration—you must twine your duration with them, for a summary would just miss the point. The durations these novels take are constitutive of their meaningfulness, and if you fully grant them their time they will reward you by restoring thick qualities to your own temporal experience.[25] Reading novels like Richardson's—novels of sensibility—fashions subjectivities alert to duration as a continuous whole saturated with a variety of feelings; in contrast, the entertainment industry fashions distracted subjectivities, whose sense of time is fragmented into discrete moments that come in no necessary order. Fragmented form makes such time easily alienable and more akin to other commodities in the market that engenders it—a moment at the theater swappable with a moment at a pleasure garden, again swappable with reading the first few pages of a novel. In Diderot's account, absorption in Richardson's novels ameliorates this condition by restoring to actions their durations, thus making any moment tied more strongly to the full process of which it partakes.

For Diderot, Richardson's novels are a palliative not only for fragmentable, alienable time, but also more generally for the alienated individualism that Diderot seems to suppose—like more recent critics of modernity—as the condition of his world. For Richardson's novels, Diderot argues, are his companions, and their companionship is peculiarly durational: "I still remember the first time I came across Richardson's work: I was in the country. How delightfully moved I was by them! With every moment I saw my time of happiness growing a page shorter. Soon I had the same feeling as is experienced by men who get on extremely well together and, having been together for a long time, are about to separate" (84). Diderot characterizes his "first time" with Richardson by his being "delightfully moved"; his happiness glossed as a dreading of time running out; his attachment identified as a long-forged familiarity soon destined to end. But even more interestingly, Diderot describes the friendship of novels as the pleasure we derive from conversing with friends. Recall that he concludes his praise by exclaiming with pathos: "Emily, Charlotte, Pamela, dear Miss Howe, as I talk with you, the years of toil and the harvest of laurels fall away, and I go forward to my last hour, abandoning every project which might also recommend me to future ages" (97). Richardson's novels divert Diderot from the pursuit of worldly achievements, but they reward him not only with a high quality and inalienable durational experience but also with companions to converse with. Novels provide Diderot with friends, and friendship is specified as the uniquely emotional duration of intimate conversation.

In *Spectator* 225 (17 November 1711) Addison comments that "the Talking with a Friend is nothing else but *thinking aloud*" (2:375). And thinking aloud, like writing essays, benefits from selection and composition, as Addison's defense of discretion throughout this essay suggests. Thus we might conclude that companionate conversation amounts to the succession of ideas in common—the primary experience of duration as it is shaped not just by compositional forms, but also more specifically by discursive forms that are inherently social. For both Addison and Diderot, at stake in compositional forms is communication—the succession of ideas not in solipsistic privacy but in social circumstances, whether in intimate conversation or in print. This social dimension of human durational experience is harder to gauge when examining Locke's and Hume's philosophies of time; however, insofar as Hume approaches duration as aesthetic form—a logic of composition that must be sensibly apprehended—he underlines the constitutive communicability of qualitative duration. For a logic of composition—a pattern—is what makes something iterable—recognizable to another person and, thus, potentially shared.[26]

Moreover, that Diderot describes novelistic characters as friends indicates how the succession of ideas in common can be conveyed through a medium that is both mechanical and fictional—novels, which communicate through print feelings of and for nonexistent persons. This serves to remind us that even as the sensibility chronotope highlights qualitative experience, it has nothing to do with singular particularity. Experiencing qualitative duration by means of reading novels, then, returns us to abstractions and evokes the technological time of chronometry and homogenous empty grids, albeit in a very different way from the one dominating Locke's discussion. Michael McKeon has pointed out that even as eighteenth-century literature invokes sensible experience, it relies on the model of scientific experiment, imagining the temporality of reading in terms of its potential for repeatability. That is, literature consciously provides for ways in which its claims can be successfully iterated, provisions that, McKeon suggests, are well articulated by what Samuel Johnson calls the "test of time": "a generalizing abstraction from variables—the personal, the local, and the temporary," in McKeon's paraphrase.[27] For Diderot, as well as for the novelists examined in the following chapters, such generalizability includes carefully devising techniques for transposing, across homogenous empty time, embodied and emotional experiences of qualitative duration. And this, in turn, suggests that qualitative duration gains its value within the sensibility chronotope not only when individuals experience it as inextricable from their own endurance, as in Locke's example of being shot or as in Diderot's experience of reading Richardson, but also when it recurs experientially across bodies and subjectivities, well after any one of its particular manifestations and individual endurances has ended. Insofar as novels convey qualitative duration to readers anywhere and at any time, they point up the fundamental sociability of qualitative duration by insisting that it is finite on the level of the individual, while being infinite on the level of the species. We find stronger articulations of this social dimension of the sensibility chronotope in analyses of durational ethics and aesthetics in Hutcheson, Smith, and Burke and in novels by Richardson, Sterne, and Radcliffe, to which I turn in the following chapters.

*Chapter 2*

# Temporal Moralities and Momentums of Plot

## Richardson and Hutcheson

In *Spectator* 225 (17 November 1711) Joseph Addison distinguishes between discretion and cunning based on their differing temporal horizons. "Cunning is a kind of Short-sightedness" (2:376), he argues, while "The Cast of Mind which is natural to a discreet Man makes him look forward into Futurity, and consider what will be his Condition millions of Ages hence, as well as what it is at the present" (2:377).[1] Futurity has been often invoked as a reference point for Christian ethics, which sought to find emblematic signs of an eternity yet to come in the present.[2] But Addison suggests that a discreet man focuses on a futurity in this world rather than in the next—a historical temporal logic that presumes causal, rather than signifying, relations between present and future. For even as our "Condition millions of Ages hence" seems hardly imaginable, Addison continues to explain that the discretion of a wise man "carries his Thoughts to the End of every Action, and considers the most distant as well as the most immediate Effects of it" (2:377). He thus advocates an ethic based on assessing what our present conditions enable and how what we do now may pan out immediately as well as further down the line. But we should also notice that in Addison's discussion it takes no less than a wise man to reckon with such horizons of judgment. Piety might encourage divining signs of God's eternal judgment in the present; cunning might enable us to compute short-sighted eventualities to our advantage; but real wisdom projects a futurity in this world that comprehends actions and their consequences in the present, in the distant future, and at each moment along the way.

Addison here neatly condenses the contours of an especially difficult ethico-temporal problem that troubled many of his contemporaries. As Lewis White Beck put it a few decades ago, "Finding *this* world quite enough is a characteristic trend of the Enlightenment, which nurtured the growth of time at the expense of eternity."[3] For many eighteenth-century thinkers, this meant recognizing that the duration within which actions transpire does not neutrally house inevitable causal chains, but rather introduces myriad destabilizing opportunities, whereby any present moment might turn out to be an occasion for the diversion of action from its intended course or an attenuation or acceleration of its momentum. Several studies have suggested that eighteenth-century moralists found such growth of time so daunting that they essentially gave up on it—either abdicating morality in favor of power, or side-stepping action by privileging deliberations and sympathy. The first tack focuses on functional means of controlling instability and argues that eighteenth-century liberalism sought to order action, rather than attempt to make it more just. The second tack highlights emotional identification and multiperspectival deliberations and argues that eighteenth-century sensibility relocated justice from the realm of action to the realms of feeling, thought, and conversation. While the ideological commitments of these analyses are very different, and while they rely on different archives—the first on records of practice, technology, and institutional developments, the second on the history of ideas—both approaches suggest that eighteenth-century moralists engaged the growth of time by devising ways to remove the variable of duration—that they focused on ordering time and on effects of spatial perspectives but not on the very temporal constitution of actions and their assessments.

In this chapter I demonstrate, by contrast, that both Francis Hutcheson and Samuel Richardson attend to the interminable instabilities of action by developing thoroughly durational models of virtue and judgment—ones that, like Addison's, advocate prospecting "to the End of every Action" and assessing "the most distant as well as the most immediate Effects of it." Hutcheson's importance in the history of moral philosophy, I explain, arises not only from his elaboration of the moral sense—his insistence that we have natural benevolent intuitions—but also from developing utilitarian principles. His utilitarian analyses underscore the temporality of action as trajectories of impact—forces that travel in time and that possess what he calls "the moment of good"—which, according to his account, must inform evaluations of both our deliberations regarding future actions and our assessments of actions that others have already initiated.[4] Though a utilitarian calculus often presumes a conception of time as a neutral and equable flow that simply houses action rather than endows it with qualities, we will see that in Hutcheson's complex discussion, duration becomes an indeterminate variable and judgment becomes a temporally qualified moment.

With Hutcheson's durational moral sentiment in mind, we can better understand Richardson's explorations of the time that extends between the perception of a situation and a decision to act, and between the instigation of an act and its subsequent consequences. This time opens up countless opportunities for actions to change their course and potential impact, as well as for revaluations of trajectories on the move. Richardson, I argue, engages these complications through the content of his heroines' recurring future projections, as well as through the form of his novels—in both his influential technique of writing to the moment and in his conception of plot as a design that requires attention throughout but that becomes fully available only on a second reading. In Richardson's comments on his technique we find an analogy between reading and judging, one that magnifies the associationist looping movement of thought that, as we saw in the previous chapter, Addison emphasizes in his *Spectator* 476. But if a periodical essay can promote in its compositional logic and demand of its readers only limited extensions of memory and anticipation, then Richardson's very long novels can mediate such exertion of consciousness over much greater expanses. And it is the magnitude of recall and, especially, future projection that turns leisure reading in Richardson's estimation (or at least the leisure reading of his novels) from an aesthetic pleasure into an ethical project. Richardson's novels, I suggest, conceive of the qualitative time of reading as an extended moment of deliberation: a timespan unified by its intense cognitive activity of weighing present possibilities against possible future trajectories and a timespan whose value arises not from the accuracy of projection—which Richardson insists can be achieved only in hindsight—but from immersion in its intense qualities.

## Richardson's Writing to the Moment and the Value of Narrative Length

Anyone who has picked up a Richardson novel knows that these demand a substantial time commitment. It takes time to read 1,600 pages, and one reason to devote such duration to reading, as we've seen Denis Diderot argue, is to remind ourselves of the sheer number of details demanding our attention when we try to get something done. "How much trouble, care and activity it takes to accomplish the smallest undertaking," to recall Diderot's remark in his "In Praise of Richardson."[5] Descriptive comprehensiveness—the empiricist attention to circumstantial detail that Ian Watt and many others highlight in Richardson and his contemporary novelists, and that Stuart Sherman places within eighteenth-century debates about whether effective representation entails exhaustive notation or critical selection—certainly explains one dimension of the bulkiness of Richardson's novels.[6] But descriptive comprehensiveness approaches duration as a means—housing or communicating details of content—rather than a value or an end in itself. In his letters, Richardson indicates a different justification for the length of his novels, one that prioritizes long and intensive durations. He suggests that in order to judge judiciously in a constitutively unstable world, we must maintain a double consciousness of both current conditions and future trajectories of action. And the record of such recurring assessments that attend to both current conditions and future projections takes time to make and takes time to read.

When Lady Bradshaigh criticizes the durational demands Richardson makes on his readers, he replies: "All I have to say for myself is, *first*, that the new Manner of Writing—to the Moment—betray'd me into it; flattering myself, that hardly any-one, who would attempt the same Manner, would be able to avoid the same Excess."[7] Richardson claims that his new compositional technique for representing immediacy of consciousness entails a combination of first-person, present-tense narration and narrative longevity. Yet the relation between the two is not at all obvious; why should first-person, present-tense reporting entail writing at length? Most scholars have not been convinced by this defense and assume no strong connection between the two hallmarks of Richardson's form. Discussions of writing to the moment have tended to consider this technique in isolation from its embedding in long narratives. There's a range of writings to the moment, Mark Kinkead-Weekes explains in what remains one of the most extensive analyses of Richardson's technique, but it runs from representing the urgency of the moment of acting to representing the urgency of the moment of response to action. And Tom Keymer, who emphasizes that there's much more to Richardson's achievement than representing immediacy, distinguishes between the novelist's epistolary method and his writing to the moment, accepting the latter as relying solely on present-tense reporting.[8] From these perspectives, Richardson's signature form involves first-person, present-tense reporting with no reference to very long narrative wholes.

But if recent scholars have not found Richardson's connecting of his writing in present tense and his lengthy narratives especially revealing, Leah Price shows that his eighteenth- and nineteenth-century abridgers insistently adhered to it, shortening his novels not only by editing out letters and scenes but also by transposing some into third-person, past-tense narration. Price pointedly observes that "a sentence phrased in the past tense and the third person is no shorter than one in the present and the first"; but even as "brevity has no intrinsic connection with narrative distance," as she puts it, Richardson inaugurates a powerful convention that associates the two as well as their inverse.[9] And yet if grammatical logic cannot help us understand Richardson's choice of combining presentness of reporting with long narrative durations, nor the compelling appeal of this combination for subsequent writers, then perhaps a logic that lies other than in language may prove more helpful. Indeed, in another letter to Lady Bradshaigh, Richardson suggests that the relation between immersive immediacy and prolixity may be grounded not in the number of words entailed by the tense and person of a sentence, but in what it takes to make first-person, present-tense reporting an effective conveyor of morals. He writes:

> Ye World is not enough used to this way of writing, to the moment. It knows not that in the minutiae lie often the unfoldings of the Story, as well as of the heart; & judges of an action undecided, as if it were absolutely decided. Nor will it easily part wth. its first impressions. How few Lady B's <that> who will read it over once for Amusement, and a second time to examine into the unjustness or justness of its several parts, as they contribute to make one Whole![10]

I return to first impressions and second readings at the end of this chapter, but at this point I want to focus on the early sentences of this passage—on Richardson's flagging of the judgment of undecided action as a different kind of judgment and on his insistence that to properly read "writing, to the moment," one must recognize a relation between present details and narrative unfolding. Readers who are not proficient in moment form, Richardson indicates, tend to make three mistakes. First, they do not recognize that a moment is not sufficient to itself; there is an openness to any given moment—"action undecided"—when it is presented from its own perspective (that is, from the perspective of first-person present-tense). Second, they fail to understand that this openness can only be settled when the moment is viewed from the perspective of the full narrative—as the "several parts . . . contribute to make one Whole." And third, they miss the small details through which a moment can be seen to carry within it seeds of the whole to which it belongs—"in the minutiae lie often the unfoldings of the Story." This last point is especially important: Richardson insists on a complex durational experience of reading: we are immersed in present detail while remaining conscious of the by-no-means certain, yet nonetheless inevitable, connection between present details and future trajectories of unfolding. This connection is necessary but also contingent, a point delivered in Richardson's qualification of the relation between minutiae and unfolding with the word "often," rather than "always." There is a necessity here of future projection from a position embedded in the present with a strong recognition of the uncertainty of such an endeavor. In Richardson's account, adequate reading of writing to the moment involves the kind of associationist prospect Addison's aesthetics highlights, one laced with an intense awareness of both the epistemological indeterminacy of such projections and our embeddedness in the present moment.

Thus the technique of writing to the moment, and, as we will soon see, the content of Richardson's stories, model for readers the possibility of conceiving of judgment as the intensely complex and time-consuming experience of judging on the move from the position of a present that worries much about the future, even if such trajectories of action can only be reliably known in hindsight—on second readings. As Richardson continues his reply to Lady Bradshaigh, he makes it amply clear that his writing seeks to model such judgments and to offer readers occasions to exercise such an ethical approach, an endeavor he takes to be both new and urgent. He writes:

> But when this hasty-judging world will be convinced that they have seen the *last work* of this too-voluminous writer, they will give what he has done, more of their Attention—*perhaps*—Especially when they are convinced that his own interest has been less his motive for writing, than that of their children in the most dangerous part of their lives. Then, madm., will be discovered and approved, if I am not a false prophet, and deceive not myself, or am not deceived by

> my vanity, those delicacies occasioned by the difficult Situations of my principal Characters, to which now they will not attend. (289–90)

People, Richardson contends, judge too quickly; they don't devote enough time and attention to evaluating his achievements, nor do they devote sufficient time and attention to assessing his characters. The two are related, for if readers were to devote a more careful and longer duration to Richardson's work, they would notice more details in his characters' circumstances and actions and recognize how much time and care it takes to assess them properly.

Why does Richardson believe his readers should be bothered with such long and intense durations of disturbingly uncertain judgments? Why does he bother to develop a technique that demands of its readers—if they do it justice—the embracing of such difficulty? Richardson, I believe, responds to one of the most urgent problems of moral philosophy of his day: the problem of time's destabilizing action and, as such, of its making judgment exceptionally difficult. Addressing the difficulty of action in time, Richardson recommends judging with duration—which is to say, making duration a key variable of our assessments in every way and at every moment. And judging with duration entails, according to Richardson, the impossibility of getting our judgments right but the increased valuation of the experience of having tried. I will say more about Richardson's proposed solution in the third section of this chapter where I consider not just his writing technique but also the content of two of his novels, *Pamela* and *Clarissa*. But before returning to Richardson, I want to examine the problems of action and judgment in more detail and provide a context for Richardson's approach in Hutcheson's moral philosophy.

## The Problem of Action and Hutcheson's Moment of Good

Eighteenth-century moralists worried about the difficulty of following actions to their destinations in an increasingly crowded and fast-paced world. Identifying the clearest articulation of this modern crisis in Bernard Mandeville's *Fable of the Bees*, Vivasvan Soni explains: "Virtue, Mandeville claimed, was an archaic survival from the small republican societies of yore. In such societies, virtuous actions had beneficial effects and contributed to the flourishing and happiness of society. In the large, trading societies of today, however, virtue is at best ineffective, and at worst a misguided code of conduct that produces social stagnation. . . . Individual action becomes impotent or loses its meaning, not because it is without effects, but because the effects of actions have become radically unpredictable."[11] Soni, among a number of other scholars, suggests that theories of moral sentiments such as Adam Smith's address these difficulties by side-stepping the temporal dimensions of actions and instead focusing on sympathetic identification, which "privilege[s] our affective response to an immediate spectacle that exceeds narrative."[12] Other scholars, in agreement that the temporal dimensions of action have been downplayed by eighteenth-century moralists, highlight multiperspectival deliberations in theories of moral sentiments. Thus Hina Nazar details Hume's and Smith's development of judgment as an intersubjective enterprise that promotes imagining other points of view and examines the ways in which Richardson and Austen underline the heart as an object that induces observation.[13] And, as I discuss in this book's introduction, James Chandler demonstrates how sentimentalists develop a unique logic of probability that relies on an imagined swapping of positions.[14]

These three studies do a superb job of analyzing the complex rationale of one especially influential strand of theories of moral sentiments. But they all rely on a theatrical model of sympathy that gained its most influential articulations in Hume's emphasis on distance in the *Enquiry Concerning Human Understanding* (1748) and the *Enquiry Concerning the Principles of Morals* (1751) and in Smith's emphasis on spectatorship in *The Theory of Moral Sentiments* (1759).[15] Distance, position, and spectatorship aim to address difficulties in earlier articulations of moral sentiments, specifically in Francis Hutcheson's extensive case for benevolent intuition, which he calls the moral sense. As Nazar explains, Hume and Smith revise Hutcheson by making the case for the indispensability of a reflective process that reviews and corrects immediate reactions; and as Chandler points out, Smith highlights self-serving biases that Hutcheson's analysis overlooks and responds by advocating techniques for counterbalancing these biases through imagining a first-person point of view toward others and a third-person point of view toward oneself.[16] According to these discussions Hutcheson, like Richardson, develops a model troubled by its commitment to a first-person perspective that can best be corrected by multiperspectivalism. And yet, it is also worth noticing that Hutcheson, like Richardson, relentlessly insists on calculations of the long-term and wide impact of action as another way of counterbalancing a narrow first-person perspective.

In the *Illustrations upon the Moral Sense*, published together with *An Essay on the Nature and Conduct of the Passions and Affections* in 1728, Hutcheson explains that "The *Non-performance of Covenants* is made *penal*, not because of their *signifying Falshoods*, as if this were the Crime in them: But it is necessary, in order to preserve *Commerce* in any Society, to *make effectual* all *Declarations of Consent to transfer Rights* by any usual *Signs*, otherwise there could be no *Certainty* in Mens Transactions" (173).[17] Hutcheson finds the greatest threat to "commerce in any society" not in lying but in time passing; he worries less about people intending to deceive and more about people changing their minds. By this logic, the law ought not to concern itself with determining intention, but rather with stabilizing the effects of time on actions, and laws of covenants do this by pinning the value of acts at their inception—the initial declaration of a promise necessitates a particular following course of action that remains indifferent to how preferences (among other circumstances) change. This legal solution addresses instability by factoring duration out of assessments of action. But when Hutcheson considers virtue, rather than jurisprudence, he not only takes volatility to be one of the most urgent problems, but also develops solutions that are thoroughly durational.

In his *An Essay on the Nature and Conduct of the Passions and Affections with Illustrations on the Moral Sense* (1728) as well as in his earlier *An Inquiry into the Original of Our Ideas of Beauty and Virtue in Two Treatises* (1725), Hutcheson elaborates and revises the moral sense initially formulated by Shaftesbury and makes the case for our having a natural disposition towards what he sometimes calls "moral good" and sometimes calls "benevolent affections."[18] Positioning himself in sharp contrast to a tradition running from the Epicureans through Hobbes and Mandeville, who posit self-love as motivating all actions, Hutcheson argues that two different kinds of motivation instinctively compel us: on the one hand, selfish pleasures, which he also calls natural good, and on the other, benevolent affections, which he also calls moral good. In a significant departure from Hobbes and Mandeville, Hutcheson makes the case for benevolent affections as independent and equally compelling sources of motivation, criteria of evaluation, and principles of sociability, and even more strongly, as overriding stimulators of our aesthetic pleasure.[19] As he explains in the *Inquiry*: "This moral Sense, either of our own Actions, or of those of others, has this in common with our other Senses, that however our Desire of Virtue may be counterballanc'd by Interest, our Sentiment or Perception of its Beauty cannot; as it certainly might be, if the only Ground of our Approbation were Views of Advantage" (94–95).

Hutcheson often argues that the moral sense operates instantaneously. The stakes of such immediacy appear most clearly when he contrasts the working of the moral sense with the direction of reason. "Notwithstanding the mighty Reason we boast of above other Animals," he writes in the *Inquiry*, "its Processes are too slow, too full of doubt and hesitation, to serve us in every Exigency, either for our own Preservation, without the external Senses, or to direct our Actions for the Good of the Whole, without this moral Sense. Nor could we be so strongly determin'd at all times to what is most conducive to either of these Ends, without these expeditious Monitors, and importunate Sollicitors" (180). In equating the external senses with the moral sense, Hutcheson suggests that at the very moment in which we perceive something we also evaluate its goodness (whether narrowly for ourselves or for a greater good); reason is too slow to lead us from perception to assessment, and, therefore, it must be that we have a moral sense (rather than, say, a moral faculty)—a quicker prompt to move us from a neutral registering of stimulation to an evaluation of it. From this perspective, the moral sense looks like a willful preempting of at least one kind of instability that time breeds: the shorter the duration spanning from perception to judgment, the fewer occasions for the object itself to change.[20]

And yet it may be useful to understand the argument for the instantaneousness of the moral sense as strategic—as less a claim for an instinctive reaction that requires no time, than a way to strengthen the case for the moral sense's universality and naturalness. Insofar as the moral sense predisposes us toward the good, it requires no actual experience for its existence (though it may require experience for developing to its fullest potential).[21] It counts, thus, as an atemporal capacity. And yet, the moral sense is nothing but an experience insofar as it operates by combining a sensation of pleasure with a judgment about an actual object that seems to have caused its activation. It is, thus, a temporal phenomenon, subject to the scrutiny of causes and effects. As Tony C. Brown points out, such a combination of atemporality and temporality is especially difficult to articulate and has been key to a number of eighteenth-century examinations. Brown looks at conjectural histories of the origins of languages and at essays on aesthetics, demonstrating that they reference the primitive and exotic whenever they reach the difficult juncture of the temporal and atemporal. Brown explains these as rhetorical moves, signaling the difficulty of thinking the two orders together.[22] Hutcheson's compression of the duration of experience when it relies on the a priori capacity of the moral sense may likewise be understood more as a rhetorical move that signals the heterogeneity of these orders, than as an argument about the duration of judgment per se.

I propose such an interpretation of the moral sense's immediacy because, as Stephen Darwall insists, Hutcheson repeatedly aligns the moral sense with calm reflection. As Hutcheson puts it in the preface to his *Inquiry*, "that Determination to be pleas'd with the Contemplation of those Affections, Actions, or Characters of rational Agents, which we call virtuous, he [the author] marks by the name of a Moral Sense" (9).[23] The moral sense is thus a feeling that transpires in thought, and thinking by its very seriality takes time. Moreover, as we will see, the species of contemplation that pleases involves for Hutcheson some very complicated assessments of the impact of actions in both time and space—assessments that cannot but take time to make.

Considering fully all the circumstances of an action from the position of calm reflection entails for Hutcheson not a narrow assessment of present circumstances, but a projection of an action's impact in social space and in time. In his preface to the *Inquiry* Hutcheson writes: "We generally acknowledge, that the Importance of any Truth is nothing else than its Moment, or Efficacy to make Men happy, or to give them the greatest and most lasting Pleasure" (7). Though this is a statement about truth, at stake here is not differentiating truth from falsehood, but ranking different truths according to their usefulness for promoting happiness. As such, it anticipates Jeremy Bentham's famous definition of utility in his 1781 *The Principles of Moral and Legislation*: "By the principle of utility is meant that principle which approves or disapproves of every action whatsoever, according to the tendency which it appears to have to augment or diminish the happiness of the party whose interest is in question: or, what is the same thing in other words, to promote or to oppose that happiness."[24] Yet, differing significantly from Bentham, Hutcheson uses the word "moment" as shorthand for utility: "Moment, or Efficacy to make Men happy, or to give them the greatest and most lasting Pleasure." This is no accidental flourish. In Treatise II of the publication that this sentence prefaces, Hutcheson devotes much attention to what he calls "the moment of good," by which he means a compound assessment through which we "compute the

Morality of any Actions, with all their Circumstances" (128), attending to "the Number of Persons to whom the Happiness shall extend" (125), the "Dignity of Persons" affected (125) as well as their "Indigence" (123), and the possibility of unintended consequences (125–26).

Why, we might ask, does Hutcheson concern himself with "the moment of good" rather than with "good" as such? In the doctrinally and culturally diverse world for which Hutcheson is writing, establishing common ground for evaluating goods is difficult enough; what does it entail for Hutcheson to insist that not only goods, but also their moments, require assessments?[25]

For Hutcheson and his first readers, "moment" invoked multiple meanings, some highly technical and resonating with the authority of the new sciences. Ephraim Chambers's *Cyclopaedia* (1728) entry for "moment" begins with "the most minute and insensible Division of Time; or what we otherwise call an *Instant*"; then proceeds to the "new Doctrine of Infinites" in which "moment" denotes "the infinite small Parts of Quantity" and is, thus, synonymous with infinitesimals; continues to the study of "Inceptives" in which "Moments are the generative principles of Magnitude"; and concludes with the study of mechanics in which "moment" "is the same with *Impetus*, or Quantity of Motion in any moving Body." Samuel Johnson's *Dictionary of the English Language* (1755) offers a more colloquial definition comprised, nonetheless, of three different meanings: "Consequence; importance; weight; value," "Force; impulsive weight; actuating power," and "An indivisible particle of time." Hutcheson's general remarks about the utilitarian calculus and his use of "moment" in his preface to the *Inquiry* suggest Johnson's first two colloquial definitions: the moment of good underlining the variability of both values of goods and the impact of their consequences, as well as the dynamic nature of any good whose source is action. Thus the "moment of good" implies attention to consequences with variable values or weights, as well as attention to intentions that have a propelling power. But Hutcheson also means to give "moment of good" scientific authority, using mathematical formulas to define it in the *Inquiry* and in the *Essay*. And close attention to these formulas, and a consideration of why he thinks we need two different ones and how we might relate their potentially different estimations, suggest the relevance of Chambers's more technical definitions—impetus of motion, generative principle, unit that might be both small and infinite, and finally, a moment as a unified temporal segment.

Hutcheson's first formula, presented in the *Inquiry*, highlights the agents of action, intentions, and beginnings: "The moral Importance of any Agent, or the Quantity of publick Good produc'd by him, is in a compound Ratio of his Benevolence and Abilitys: or (by substituting the initial Letters for the Words, as M = Moment of Good, and μ = Moment of Evil) M = B × A" (128).[26] In the *Essay*, however, the formula focuses not on acting agents, but on actions' effects, and these effects are crucially determined by duration: "In computing the *Quantities* of Good or Evil, which we pursue or shun, either for our selves or others, when the *Durations* are equal, the Moment is as the *Intenseness*: and when the *Intenseness* of Pleasure is the same, or equal, the Moment is as the *Duration*. Hence the *Moment* of Good in any Object, is in a compound Proportion of the *Duration* and *Intenseness*" (37). He continues in the *Essay*: "To an *immortal* Nature it is indifferent in what part of its Duration it enjoys a Good limited in Duration. . . . But if the Duration of the Good be *infinite*, the Earliness of Commencement increases the Moment, as *finite* added to *infinite*, surpasses *infinite* alone. To Beings of *limited certain Duration*, Axiom 12 [the preceding sentence] may be applied, when the *Duration* of the Good would not surpass the Existence of the Possessor, after the Time of its Commencement. To Beings of *limited uncertain Duration*, the Earliness of Commencement increases the Moment of any Good, according to the Hazard of the *Possessor's Duration*" (38).

These two formulas together indicate that the moment of good generates action in subjects that has impact on objects. It is, thus, a travelling force. And yet Hutcheson insists on estimating this traveling force at both its inception and end, which suggests that the assessment of benevolence and ability in an agent—the moment of good as defined in the *Inquiry*—may yield a different quotient of good from the evaluation of intensity and duration of her action's impact—the moment of good as defined in the *Essay*. A ball will not hit its target with as much power and the exact same directionality as the thrower invested in it. There's friction as well as gravity, and there might be wind and other balls in the air that may interfere with its course. Similarly, if Pamela's success appeared so improbable to Henry Fielding, this is because he would not expect that a person of such meager abilities should have such a durable and intense impact. And if Clarissa's death seemed a disputable outcome to Lady Bradshaigh, then this is because a heroine of such stature and benevolence could be expected to have a more positive impact. In other words, that Hutcheson develops two formulas points out that neither one is sufficient for assessing good—that indeed the "moment of good" is borne out neither by acting agents through their benevolence and ability, nor by the effects of action, intended or not. If we usually understand utilitarian calculations to presume that estimations of an action's trajectory are not impacted by the position from which they are made, Hutcheson's insistence on two different formulas for an action's inception and its end suggests that at least the temporal (if not also spatial) position from which we judge qualifies our judgment. Moreover, Hutcheson's recognition of the indeterminacy of trajectories of action indicates that the temporal medium within which actions travel cannot be equable and neutral. As we've seen in the previous chapter, about a decade after Hutcheson published his *Inquiry* and *Essay*, Hume argued much more directly that our vantage point in time determines our feelings about other points in time; but Hutcheson's conception of utilitarian estimations as requiring varying assessments according to their temporal positions also suggests, albeit implicitly, that varying temporal segments are incommensurable with one another.

Since neither the beginning of action nor its end, according to Hutcheson, is fully adequate to the task of judgment, some kind of combination between the two, or some kind of relation between the two, must be established. But what exactly is this relation? In the *Illustrations* Hutcheson argues that "the *Tendency* of our own Actions we cannot always know; but we may know certainly that we *heartily* and *sincerely* studied to act according to what, by all the Evidence now in our Power to obtain, appeared as most *probably tending to publick Good*. When we are conscious of this *sincere Endeavour*, the *evil Consequences* which we could not have foreseen, never will make us *condemn* our Conduct. But without this *sincere Endeavour*, we may often approve at *present* what we shall *afterwards* condemn" (175–76). If prospection of the consequences of action is thorough, then it is privileged over actual consequences in judgments of virtue.[27] Here Hutcheson highlights deliberations that precede action, much like the strand of moral sentiments that Chandler and Nazar study, but he emphasizes that the relevant imaginative projection involves estimating not other points of view in space, but changes in time—a forecasting of the trajectory and impact of an action after its travelling has already begun. Also note that this is a way of looping the computation of the moment of good, as it pertains to impact, back onto the capacities of agents: if deliberation is long and thorough enough, then this becomes a way to assess the agent's benevolence and ability.

By such an interpretation, the instantaneousness of the moral sense does not imply a short duration. Rather, it implies "an indivisible particle of time," to recall Johnson's third definition—all its duration tending towards estimating the trajectory of a possible act—as well as infinity, to recall Chambers's second definition—its potential for endless internal growth, even as it is a circumscribed unit—"the Infinite small parts" of that "Quantity" we have called "the moment of good." Thus Hutcheson's attention to the indeterminacy of trajectories of action also highlights the very moment of judgment itself as an especially peculiar, and difficult, exertion. If assessments look different at inception and at end, then neither one yields reliable knowledge of the action; both, however, are weighted with the importance of the endeavor of estimating by way of projection. In other words, although in his arguments and propositions Hutcheson aims to give us tools for judging accurately, the proliferation of these tools highlights the uncertainty of getting judgment right, while increasing the value of the very activity of judgment, not for its truth but for its effort—the "*sincere Endeavour*" as he puts in the *Illustrations*.

By founding his theory of virtue on the heterogeneous elements of moral sense and utilitarian calculus and by anchoring his utilitarian calculus in the heterogeneous assessments of beginnings and ends, Hutcheson underlines the importance of continuing to attend to duration and action, even in a populous and moving world in which carrying an action to its intended end is nearly impossible. I now want to return to Richardson in order to demonstrate how he further develops this approach. Richardson clarifies the relation between the extensity of prospection and the present moment of its undertaking and introduces periodic reassessments of the moment of good *in medias res* as a way to provide for the disparities between the beginnings and ends of action.

## Judging on the Move in *Pamela* and *Clarissa*

For decades, scholarship has by and large accepted Ian Watt's account of Richardson as mastering a realism of presentation rather than a realism of judgment (in which his rival Fielding excelled, according to Watt) and as inaugurating an influential tradition of psychological realism, thus privileging consciousness over action.[28] Recently, two studies have unsettled this consensus, making the case for *Clarissa* as a profound meditation on action—Jonathan Kramnick's *Actions and Objects*, which argues that in the final analysis Richardson privileges inaction, and Sandra Macpherson's *Harm's Way*, which argues, to the contrary, that Richardson traces actions to their very ends, to the extent of promoting an ethic of accountability for unintended consequences.[29] These analyses help us better understand the importance of action for Richardson and for novelistic realism more generally. And yet, if we continue to agree that realist fiction, and Richardson's in particular, focuses on the everyday, the mundane, the ordinary, then one thing this entails is a concern about how we judge from the midst of action and at every moment. Macpherson, in taking up the legal perspective of strict liability, imagines judgment from the position where action has already reached its ultimate destination. This position is extraordinary in the sense that rarely in our lives do we reach such a full view of an action's arc. Strictly speaking, it's a once-in-a-lifetime possibility; a little less strictly, it can happen whenever our acts are brought before a court of law, an occasion which necessarily imagines these acts as sufficiently complete—but such an event for many of us is even rarer than death. Kramnick, in taking up the perspective of philosophy of mind, imagines judgment from the position that precedes action. This position is as extraordinary as the one imagined by laws of strict liability, in that rarely in our lives can we deliberate in stillness for as long as we want. In ordinary circumstances we need to judge while in motion—while both we and our actions are on the move—despite our inability to know their final destinations or our inability to feel as though we've fully initiated exactly what we wanted to have done. And while this trivializes judgment—indeed, makes it mundane—Richardson's novels are nothing if not detailed records of the ambition and difficulty of such countless judgments on the move.

Or as Clarissa puts it, "Who can command or foresee events? To act up to our best judgments at the time is all we can do" (105).[30] It is important to note that Clarissa emphasizes that we act and that action ought to be informed by judgment. But what constitute "our best judgments"? As I argue elsewhere, and as Nazar further confirms, in *Clarissa* (1747–48) Richardson specifies justification as multiperspectival deliberative processes.[31] Here I want to suggest that these deliberative processes do not only promote multiple perspectives in space; no less insistently, they advocate projections of future eventualities. And here we can begin to recognize a Hutchesonian attitude toward the growth of time—one that tries to imagine what judgment might look like from the vantage point of a time that is as irrevocable as it is unforeseeable. Richardson's commitment to action and consequences resonates with a

Hutchesonian moral sentiment in which someone's moral sense is established through the extent to which she does her best to estimate the moment of good of all events that might be traceable to her.

Both Pamela and Clarissa spend much energy and time trying to estimate the moment of good of possible actions—how much pleasure or pain, innocence or doom, they are likely to spread in both space and time. Take, for example, an early conversation in *Pamela* (1740) in which Pamela declines Mr. B's request to visit her at Lincolnshire. Although he holds the keys to her release, she believes that once he arrives on the premises only a single narrative trajectory remains possible: B will impose sexual submission leading to ongoing doom. But Pamela refrains from explicitly naming the sexual drama and instead elaborates its temporal structure with a projective calculus that sounds very similar to Hutcheson's formula for the moment of good in its estimations of impact: "An Hour of Innocence is worth an Age of Guilt; and were my Life to be made ever so miserable by it, I should never forgive myself, if I were not to lengthen out to the longest Minute my happy Time of Honesty" (137).[32] If Pamela invites B, she hurries the onset of guilt; if she declines to extend an invitation, this may only postpone his arrival—after all, B doesn't need her invitation to visit his own estate—but this extends the duration of innocence, even if only for an hour or for "the longest minute."

Pamela imagines the trajectory of an act as fully foreseeable and the value of its impact as easy to calculate; there is no heterogeneity of happiness, no multiple possible sequences of events. Such an approach not only contradicts Clarissa's lament of the difficulty of gauging the future, but also makes judgment seem mechanical rather than deliberate. But if Pamela learns anything through her trials, it is that actions can be diverted, be revoked, or turn out to have different meanings from what they initially seemed to. Life in Lincolnshire without B's presence may preserve Pamela's innocence, but she also comes to recognize that it is very unhappy. As she reasons herself out of suicide, she comes to think that perhaps B's arrival may improve her circumstances, rather than inevitably promote her demise: "How knowest thou, tho' the Prospect be all dark to thy short-sighted Eye, what God may do for thee, even when all human Means fail? . . . And who knows, but that the very Presence I so much dread, of my angry and designing Master, (for he has had me in his Power before, and yet I have escap'd) may be better for me, than these persecuting Emissaries of his, who, for his Money, are true to their wicked Trust, and are harden'd by that, and a long Habit of Wickedness, against

Compunction of Heart?" (173). Though Pamela relies on providence, rather than on worldly instability—divine intervention could divert B's actions, rather than the conditions of action in this world making effects less than fully predictable—she combines the reference to providence with the discourse of moral sentiments: Jewkes and Colbrand (B's "persecuting Emissaries") are faulted for prioritizing private interest ("Money") over sympathetic good ("Compunction of Heart").

Moreover, upon B's arrival Pamela learns that even without divine intervention, people may hold different estimations of actions, both about the likelihood of varying trajectories and about the goodness of particular sequences. When B arrives uninvited, he neither desists from trying Pamela nor imposes himself by force; instead, he offers her a mistress contract. B proposes material riches for Pamela and her family in exchange for "Twelve-month's Cohabitation," after which, he indicates, marriage may follow: "If my Love increases for you, as it has done for many Months past, it will be impossible for me to deny you any thing" (191). To B's sex-may-lead-to-marriage projection, Pamela responds by arguing that sex does not increase love, but ends it. In any case, she adds, such sequencing can never be good: "Give me Leave to say, Sir, in Answer to what you hint, That you may, in a Twelvemonth's time, marry me, on the Continuance of my good Behaviour; that this weighs less with me, if possible, than any thing else you have said. For, in the first Place, there is an End of all Merit, and all good Behaviour, on my Side, if I have now any, the Moment I consent to your Proposals" (192). Pamela's reply confirms the temporal logic of love as passion that Niklas Luhmann has laid out in his analysis of this code. But Pamela articulates this logic in the vocabulary not of racy seventeenth-century French fiction, but of Hutchesonian morality.[33]

B's arrival in Lincolnshire doesn't inaugurate an inexorable motion to rape and an age of guilt. Instead, it functions as just one more twist in a plot with many turns. When Pamela finally embarks on her long-awaited journey to her parents, she does so neither as a violated fallen woman, nor as a triumphantly virtuous lady: "So away drove the Chariot! And when I had got out of the Elm-walk, and into the great Road, I could hardly think but I was in a Dream all the Time. A few Hours before in my Master's Arms almost, with twenty kind Things said to me, and a generous Concern for the Misfortunes he had brought upon me; and only by one rash half Word exasperated against me, and turn'd out of Doors, at an Hour's Warning; and all his Kindness changed to Hate! And I now, from Three o'Clock to Five, several Miles off"

(245). Very soon this latest twist, by which release from imprisonment has turned into banishment, will also be reversed; if on Sunday afternoon Mr. B orders Pamela to leave his presence forever, by Monday morning he sends her a letter declaring his enduring love and begging for her return. Indeed, volatility, unpredictability, actions revoked and changing course—these are the order of the day in Richardson's novels.

But this volatility seems to happen to Pamela, and while she is committed to prospection—a Hutchesonian projection of actions' moments of goods—she doesn't preempt, nor even seem aware of, how duration interferes with her projections. Clarissa, as I suggest above, is much more aware of the difficulties of forecasts, and in his second novel Richardson makes a much more direct case for the necessity of prospection in the face of inevitable instability. Repeatedly as soon as Clarissa acts, she loses confidence in her choices, recognizing, on the one hand, that "I see not how I could do otherwise than I have done" (352), while admitting, on the other hand, that "when I found it gone, I began (as yesterday morning) to wish it had not: for no other reason, I believe, than because it was out of my power" (352). But despite these worries, Clarissa does not give up on future projections. On the contrary, these difficulties press on her the necessity for periodic reassessments and flexibility. When she decides to accept Lovelace's assistance in fleeing Harlowe Place—because, as she explains, "I have no way to avoid the determined resolution of my friends in behalf of Mr. Solmes; but by abandoning this house by his [Lovelace's] assistance" (350)—she is acutely conscious of how other people's responses can divert the course of her intentions, and thus she forecasts several possible scenarios. Clarissa warns Lovelace that once she leaves her home without her family's permission she is likely to be denied her estate, but that she will neither sue her family nor marry while destitute. She adds, however, that if that happens she promises that she "will never be any other man's while he remains single or is living" (350). She also explains that she prefers not to go to the homes of his family members, so that "it may not appear to the world that I have refuged myself in his [Lovelace's] family" (350), and yet "if I find myself in danger of being discovered and carried back by violence, I will then throw myself directly into the protection of either of his aunts who will receive me" (350). Such an awareness of how time harbors opportunities for destabilizing her best-projected choices justifies in her opinion what might seem to others as fickleness; she must have the flexibility to change her mind and thus to readjust the course of her actions in the face of other people's moves. She explains this much to Lovelace when she asks him to check her news as frequently as he can, because her family's maneuvers may demand modifications of her original plan to leave Harlowe Place with his assistance. She writes: "That their [the Harlowes'] suspicions seeming to increase, I advise him [Lovelace] to contrive to send or come to the usual place, as frequently as possible, in the interval of time till Monday morning ten or eleven o'clock; as something may possibly happen to make me alter my mind" (351).

Clarissa's prospections focus on the possibility of an eventual reconciliation with her family, emphasizing how she would have to reassess her situation and choices given their responses; Anna's prospections, in contrast, focus on Lovelace and on public opinion. Anna forecasts that neither Lovelace's virtue nor the generosity of the public can be relied on. She pleads, "One thing, in your present situation and prospects, let me advise: it is this, that if you *do* go away with Mr Lovelace, you take the first opportunity to permit the ceremony to pass" (354), and reiterates shortly after "considering Lovelace's character, I repeat my opinion that your reputation in the eye of the world requires that no delay be made in this point, when once you are in his power" (355). Anna's prospections turn out to be more pertinent than Clarissa's as the novel continues to develop with Lovelace relentlessly deceiving Clarissa, imprisoning her, and finally raping her. These actions not only wildly divert Clarissa's from their intended course, but also create conditions that all but cancel the little control over the trajectories of her actions that she might have had to begin with. And yet even in such circumstances, Clarissa insists on attempting some forecast. "All my prospects are shut in" (901), she laments after being raped; nonetheless, she demands of Lovelace, "But tell me (for no doubt thou hast *some* scheme to pursue), tell me, since I am a prisoner as I find, in the vilest of houses, and have not a friend to protect or save me, what thou intendest shall become of the remnant of a life not worth the keeping?" (900). If she cannot act by the guidance of her own assessments of a moment of good, then she demands that at least Lovelace undertake and disclose such prospection.

Moreover, Clarissa values future projection to the extent of making it her final action and her primary preoccupation during the remainder of the novel. Once Clarissa collects herself from the initial despair and rage over what has been done to her, she turns to writing a final will—a remarkable act for a person as exhausted, isolated, and destitute as she is. She has very little property to distribute and few people on whom she can count to accept her belongings or her wishes. There is thus little material good to spread by means of this will. And yet in exerting herself in this way, Clarissa demonstrates how to properly take up even death as a willed act—willed not in the sense of causing her own death, but in the sense of having consequences for which it is her duty to provide. If we are to accord any significance to this document to which Clarissa devotes her last days and her waning powers, then we must recognize the thoroughness with which it realizes Clarissa's durational ethic. For it is in the writing of her will, even more than on other occasions, that we see that "to act to our best judgments at the time" means for Clarissa not so much to initiate a trajectory of events, nor to take blame for all that happens, but to project as best she can how events that are already in motion might pan out, before they actually make their impact.

In the preamble to her will, Clarissa explains that "I for some time past employed myself in penning down heads of such a disposition; which, as reasons offered, I have altered and added to; so that I never was absolutely destitute of a *will*, had I been taken off ever so suddenly. These minutes and imperfect sketches enabled me, as God has graciously given me time and sedateness, to digest them into the form in which they appear" (1412). Indeed, being "taken off ever so suddenly" is the story of her life—first, the unexpected fortune that her grandfather bequeaths her, then the unexpected malice that her siblings promote against her, then Lovelace's whisking her off at the gate of Harlowe Place despite her remonstrances, and finally several occasions in which she believes she manages to break free from him only to be taken over by him again. But in the preamble to her will, Clarissa emphasizes that regardless of when she might have passed away from this world, she would not have been "absolutely destitute of a will." Whatever stage she achieved in her deliberations would count as her intentions, for intentions are always imperfect; and although here—for once—she was given sufficient duration and calm, her will remains comprised of "imperfect sketches."

If the very writing of a will in Clarissa's case is not remarkable enough, its content is astounding in the plurality of possible trajectories it imagines and seeks to provide for. If Mr. Harlowe accepts his daughter's body, then she is to be buried in the family vault; but if Mr. Harlowe prefers to disown her, then she is to be buried in the churchyard of Mr. Smith's parish. If Lovelace does not impose himself, then her body is not to be viewed by anyone outside her close family; but if he continues to insist, then he may view her provided that he also read a cautionary note. If Mrs. Harlowe should like to keep her daughter's picture, then Clarissa bequeaths her the one that used to hang in her parlor; but if she prefers not to keep it, then it goes to Aunt Hervey. This will is thus a virtuoso example of Clarissa's Hutchesonian virtue: in it, she projects her moment of good like an expert juggler, rewriting her death as a fiat of keeping multiple balls flying in the air.

Through Clarissa's will, Richardson makes his most programmatic case for prospection and durational judgment. Wills provide an institutional form to the Hutchesonian moral sense, giving legal standing to a virtue constituted by planning for the future. Two other legal institutions cast ominous shadows in this novel—marriage and the courts. Eighteenth-century English marriage contracts relied on pinning judgment on beginnings and intentions; a contract undertaken before cohabitation and enduring for a lifetime, marriage entailed accepting a commitment to partnership that would remain indifferent to changing circumstances and desires. Clarissa, tellingly, repeatedly rejects this approach to binding legal terms and pays for this dearly, though the novel justifies her choice. She refuses many suitors, even before Solmes and Lovelace, we are told, and often expresses her intention never to marry. But Clarissa also repeatedly avoids courts—that legal institution that makes agents retroactively accountable to the full trajectories of their actions. In both the beginning and the end of the novel she refuses to press charges, first opposing Anna's advice that she sue her father to regain control over the property her grandfather bequeathed her, and later rejecting Dr. Lewen's advice that she file a rape suit against Lovelace. Wills, on the other hand, neither focus on the beginnings of actions nor on their endings, but instead highlight the importance of planning for the future, and in Clarissa's case, emphasize how this planning might remain attuned to the duration in which it transpires via provisions for flexibility and reassessments.

But in *Clarissa*, unlike in *Pamela*, Richardson also emphasizes that this durational ethic suffers competition. While in *Pamela* the two lead characters share a similar temporal conception, Clarissa and Lovelace clash in their very chronotopes. Both Pamela and B inhabit, for the most part, a historicist diurnal paradigm. Even as Pamela occasionally invokes Christian emblematic time, calendar and clock underwrite both hers and Mr. B's thinking and action. Thus Pamela and B incessantly negotiate schedules. When B makes Pamela his first offer, for example, he asks her to consent not to the terms of the deal but to his proposed time frame: "Now, *Pamela*, when I have stoop'd so low as to acknowledge all this, oblige me only to stay another Week or

Fortnight, to give me Time to bring about some certain Affairs" (84). And Pamela obliges him at least in this way of framing the discussion, for she does not negotiate money or arrangements for the Andrews family, only the schedule for deliberations. "Sir, said I, I beg at least two Hours to consider of this," Pamela asks, and B replies: "I shall, said he, be gone out in one Hour, and I would have you write to your Father, what I propose." Pamela agrees: "Sir, said I, I will let you know in one Hour then my Resolution" (87). Similarly, when Pamela and B finally reach agreement about their mutual attraction, they continue to negotiate the scheduling of their wedding. Pamela wants to marry on a Thursday, while B argues that he cannot wait seven days longer and proposes Monday instead. This scene makes clear that both value timing and fundamentally agree on the calendar as providing the appropriate terms for discussions of time.[34]

In *Clarissa*, by contrast, we find a struggle not over scheduling within an agreed framework of continuous linear time, but far more irreconcilably, over the very nature of temporal experience and temporal order. Clarissa and Lovelace's negotiations of timing involve asking whether the present gains its value in relation to the growth of time or in relation to eternity. The latter framing might recall the Christian emblematic conception that I mentioned in my introductory comments to this chapter, but Richardson's presentation installs a startling reversal: it is the libertine, rather than the devout Christian, who advocates the horizon of eternity, while the virtuous insists on the growth of time.[35] In fact, Richardson doesn't seem to worry much about the competition between a Christian chronotope and an historicist one. Unlike later gothic writers such as James Hogg for whom emblematic justice runs the grave risk of condoning earthly crimes, for Richardson it seems easy to reconcile Clarissa's worldly morality with her piety. Much more worrying for him is the competition with the kind of sentimental ethic that privileges sympathetic immersion in the present at the expense of judgment of trajectories of action.

Lovelace quite obviously instantiates a passionate spectatorship that reveals how perilously close sympathetic observation can be to erotic gazing. As he observes Clarissa's anguish after he has raped her, he cannot but be affected sympathetically. As he puts it, "I had thought I could have gazed the dear creature into confusion—but it is plain that the sense she has of her wrongs sets this matchless woman *above all lesser, all weaker* considerations" (900).

Instead of confusing Clarissa, what Lovelace observes confuses himself: "What—what-a—what—has been done—I, I, I—cannot but say—must own—must confess—hem—hem—is not right—is not what should have been—But-a—but—but—I am truly—truly—sorry for it" (901)—thus the libertine finds himself stumbling into an ethical attitude. And yet as Lovelace is pushed into remorsefulness, his sympathetic observations increasingly convey his sexual perception of his object—noticing "her enchanting neck" (913), her "bosom heaving and throbbing" (914), and calling Clarissa a "soul-harrowing beauty" (913). But Richardson in this passage criticizes not only the sexualization of sympathy; more important for my purposes, he criticizes the temporal conception underwriting such a sympathetic ethic. As Soni has pointed out, this ethic relies on tableaus of presence that are essentially severed from narrative trajectories, a point Richardson makes when Lovelace confesses: "When I am *from* her, I cannot still help hesitating about marriage, and I even frequently resolve against it; and am resolved to press my favourite scheme for cohabitation. But when I am *with* her, I am ready to say, to swear, and to do, whatever I think will be most acceptable to her" (915). Lovelace's desire to do right by Clarissa arises in the occasion of presence and can little be relied upon beyond the moment of encounter.

Richardson puts this temporal conception into competition with Clarissa's durational ethic, which relies on temporal continuity and prospection. Throughout the elopement scene—and throughout the novel more generally—two conflicting temporal attitudes struggle to shape judgment and action: Clarissa's, which is committed to the growth of time, and Lovelace's, promoting the collapse of time—a now/never binary with an ever-diminishing mediating duration in between. As Clarissa hands Lovelace the letter he failed to pick up before their scheduled elopement from Harlowe Place and asks him to leave without her, Lovelace protests: "To leave you now is to lose you for ever!" (374). He continues: "Do you hesitate a moment?—Now is the time" (375), "You never, never can have such another opportunity" (376), and "The way is clear; just now it is clear!—but you may be prevented in a moment" (376). Even more, "if you tarry you will inevitably be Solmes's wife" (375), he warns, to which Clarissa replies, "I shall gain time at least—I am sure I shall—I have several ways to gain time" (375). And Clarissa continues to insist, "I was sure, I said, of procuring a delay at least. Many ways I had to procure delay" (378-79). But Lovelace is skeptical of this strategy, not only because he doesn't think Clarissa can successfully delay her family's plans for her, but also because the goal itself seems to him useless: "And what, madam, will gaining time do?" (375), he asks.

Lovelace associates empowerment with temporal condensation. His ability to control Clarissa relies precisely on his success in whisking her as quickly as possible from the gate of Harlowe place. Similarly, after installing Clarissa at Sinclair's brothel, Lovelace boasts: "All will be over in three weeks, or bad will be my luck!—Who knows but in three days!—Have I not carried that great point of making her pass for my wife to the people below? And that other great one of fixing myself here night and day?—What lady ever escaped me, that lodged under one roof with me?—The house too, THE house; the people, people after my own heart: her servants Will and Dorcas both my servants. *Three days* did I say! Pho! pho!—*Three hours!*" (541). Lovelace's increasing confidence—growing with every sentence in this passage—manifests itself through the successively decreasing duration he projects for his exertions—from three weeks, to three days, to three hours. But what's even more remarkable about his assessment is how he imagines this duration—however short he can make it—as disconnected from anything that may follow—"All will be over," he begins. Accordingly, Lovelace's report of the rape—which takes longer exertions than he has projected—consists of announcing, "The affair is over" (883). Insofar as a moment might be disconnected from all that precedes and follows it, nothing that may follow can be said to essentially belong to it. The moment gains its value through resonating with other moments such as itself—the sexual confirmation that the libertine aims to achieve over and over again—rather than through the way it sets off a chain of consequences and the way it addresses requests, prohibitions, and preconditions. Belford points out this second framework in his response to Lovelace's rape announcement: "What work hast thou made in *one guilty hour,* for a *whole age* of repentance!" (883).

When Lovelace and several of Clarissa's devoted friends urge her to marry as a way of rectifying rape, she dismisses this solution as temporally inconceivable. Responding to Lovelace's post-rape marriage proposal, Clarissa asks: "Canst thou blot out the past week? *Several* weeks past, I should say; ever since I have been with thee? Canst thou call back time?" (908). These are not rhetorical questions; rather, Clarissa points out the two competing conceptions of time that structure the struggle between Lovelace and herself. Believing that marriage can supersede rape amounts to believing that the sequence of moments doesn't matter; sex before or after vows amounts to the same thing, because sex means union regardless of its timing. In *Clarissa*'s world many subscribe to this position—all those who advocate marriage as a remedy for rape—even if their temporal attitudes rely on differing ethical frameworks (Christian or libertine). And yet for Clarissa the sequence of action matters greatly—one sequence counts as matrimonial bliss while the other counts as harm, which might be compensated but cannot be undone, because the moment in which it was done cannot be revoked. If the growth of time matters at all, Clarissa insists, then the sequences of our actions within this duration must matter as well.

Some time ago, William Warner argued that both Clarissa and Lovelace, as well as Richardson himself in his response to Lady Bradshaigh's requests for a different ending, aim to fully control temporal uncertainty, increasingly revising proposals—moments rooted in the present and an open future—into emplotted narratives in which a single moment predetermines all interpretations of the past and all possibilities for the future. Proposals, according to Warner, are occasions of active risk-taking open to the future, and the move in time invariably amounts to a reactive "stabilizing temporality of presence," as he puts it.[36] What this means, according to Warner, is that people can choose between, on the one hand, a liberatory temporal ethic of play, and, on the other, an authoritarian temporal ethic of determinism. Warner is right that some options expire in the progress of action, and he is also right that Richardson seeks some stabilization for a world in which volatility is the order of the day. And yet, I have been arguing, stabilization and a commitment to projecting consequences inevitably keeps multiple potential trajectories open—if only because, as we've seen, different agents act at cross purposes. Moreover, openness may be play for one party—the libertine who prefers to avoid the stabilization of marriage—and tyranny for another—the woman who bears the consequences in a society that considers sex without marriage as sin. And, finally, at stake is a difference that runs much deeper than proposal scenes; at stake is a competition between different comprehensive ethical-temporal conceptions—one promoting emotional identification in the present, the other insisting on the importance of trajectories of action and their forecasts along with acute attention to present conditions.[37]

But I should also emphasize that such insistence on sequential order seems to me very different from the preoccupation with discipline highlighted by Michel Foucault. The brilliance of Foucault's analysis rests in explaining how discipline controls by bypassing consciousness; one need not worry about convictions and preferences if one can govern the minutiae of movement.[38] Richardson, in contrast, profoundly cares about the ways in which what people believe about time determines their understanding of how they can act and their choices regarding what they do to and with each other, despite all the spatiotemporal constrictions imposed on them. After all, neither Pamela nor Clarissa becomes a docile body by any scheduling scheme or any attempt to limit either's movements. Clarissa succumbs only to drugs, not to temporal manipulations and spatial imprisonment, and Richardson emphasizes that such making docile of her body troubles her mind greatly but by no means makes her consciousness irrelevant. Her mad papers and subsequent escapes from Lovelace's custody, as well as her insistence that marriage cannot remedy rape, are nothing but testimonials that she will never go gently even as her body becomes ever so weak. Foucault tracks the rise of a chronometric chronotope, one that privileges linear sequencing within a comprehensive worldview that relies on mechanical models whose aim, above all, is to quantify. By contrast, my analysis of Richardson's and Hutcheson's writings documents a dimension of a sensibility chronotope for which linear sequencing is key, but only within a comprehensive worldview that seeks to assess qualities, rather than quantities, of sensible experience. I thus agree with Stuart Sherman that Richardson's novels don't quite tell the story that Foucault has told about how the partitioning and micromanagement of time and space are conducive to producing docile bodies, but my analysis highlights different reasons. Accepting mechanical and quantifying approaches as the dominant shapers of time-consciousness, Sherman aims to document a corrective history about the timing of these measurements' appropriation as "a central tool of power, a device indispensable in the 'discipline' of large multitudes (workers, students, soldiers, prisoners)." Sherman reclaims "the temporal textures and complexities of that middle period when precise clocks and watches, conspicuous but not yet ubiquitous, proffered to their owners a temporality closely calibrated, but not yet controlling."[39] I have been arguing, however, that if calendars and clocks do not control Clarissa, this is because she is committed not to order as such, but to imagining plausible narrative strings as an ethical activity. Clarissa's commitment to the linear and causal unfolding of action accommodates not just a historicist consciousness, but also a sensible one insofar as what is at stake for Richardson is the imaginative projection of future trajectories of action along with an acute sensitivity to the duration within and by which such associationist mental activity is undertaken.

## Rereading for the Plot

Richardson's novels, then, focus on action, and *Clarissa* makes an especially strong case for a durational ethic that prioritizes both the order of action and attempts to forecast its unfolding. Another way of putting it is to say that for Richardson, plot matters greatly. And yet such recasting of my argument into narratological terms requires some clarification. Richardson certainly does not want his readers to read his novels as though they are emplotted. If the very constitution of emplotment rides on what Jesse Molesworth, following Jorge Luis Borges, calls "the magic of teleology," Richardson aligns his ethical insistence on prospection with the impossibility of getting it right—at least not on a first reading.[40] To recall the letter to Lady Bradshaigh I discussed earlier, a reader makes an unfortunate mistake when she "judges of an action undecided, as if it were absolutely decided," and only those who "read it . . . a second time" can accurately discover "the unjustness or justness of its several parts, as they contribute to make one Whole." For Richardson, beginnings and middles of plots foreshadow their endings only in hindsight. But if for Richardson middles seem important on their own terms, this is hardly for the same reasons narratologists tend to specify. Richardson does not dilate his narratives with the kinds of pauses that Gérard Genette has identified as essential to narrative rhythm—arrests from action that foreground descriptions of setting (descriptive pauses) or perceptions of setting and general narratorial meditations (reflective pauses).[41] Richardson's pauses, in a different manner, highlight action; it is characters' prospections of possible trajectories of action, rather than details of setting or general thoughts, that slow down his narratives and make their middles so bulky. Nor is Richardson's plotting digressive in the way that, say, *Tristram Shandy*'s is in its interweaving different plot lines or moving the action backward and sideways as often as moving it forward.[42] For Richardson, I have been arguing, it is not only moving forward which is key, but also a self-conscious moral commitment to moving forward. And, finally, it would also be inaccurate to describe Richardson's plots as suspenseful. If, as Caroline Levine explains, "the possibility of suspenseful uncertainty emerges when the text clearly signals that it is holding something back. . . . the suspenseful text flaunts the fact that it is withholding information," then Richardson emphasizes, to the contrary, that the kind of uncertainty his plots ride on arises from the inaccessibility of full knowledge to anyone who encounters it for the first time—it being nobody's prerogative to hold back or withhold anything.[43] Nor does Richardson exactly urge his readers to suspend judgment because of such uncertainty; rather, as I have suggested in the first section of this chapter, he urges his readers—like his heroines—to judge while remaining fully conscious of the tentative nature of judgment and the necessity of recurring reassessments.

Plot matters greatly to Richardson, and even more specifically the middles of his plots matter greatly because he is developing a durational ethic founded on thinking about action. It is a digressive mode insofar as it suspends the main action, but it is a peculiar digressiveness insofar as it suspends this action for the purpose of projecting its multiple possible trajectories, and through this mental activity it insists on one's accounting for all adventures in one's life story. I am suggesting, then, that Richardson's peculiar commitment to plot relies on a refiguration of romance, transforming romance digressions from free-standing adventures to ones grafted onto characters' lives by virtue of his novels' ethical program. Moreover, I am suggesting that insofar as Richardson insists that these projections of actions can never accurately predict the movement of plot, he also underlines the impossibility of the epistemological quests that underwrite romance plots and the errors that constitute their digressions.[44] This impossibility arises from the modern conditions of especially crowded and fast-paced fields of action. But Richardson does not aim to dispense with such quests; rather, he aims to highlight the phenomenology of judgment in these modern conditions—to focus on what it feels like to imagine a path within an unpredictable future and the ethical importance of inhabiting such difficult moments of projection.

Richardson promotes a stretching of consciousness so that thinking ideally encompasses the duration of plot, at the same time that he also recognizes this to be impossible and thus dis-identifies such thinking from the actual panning out of the trajectories of action. Richardson's approach to plot might be called a peculiar species of teleology, one that constitutes it as an ideal to aspire to in our thinking, not a fact to actualize in reality. It also constitutes teleology as something that becomes apparent not through reading, but only through rereading, and, moreover, it counts rereading as an ethical exercise that simulates teleology's satisfaction but not its real-life achievement. But specifying thinking about action in this way poses a problem both for plots and for ethics. An ethic that we all fall short of is difficult to promote, and plots that can only chronicle the failure of their heroes and readers at prospection will inevitably seem implausible—belying the mainstay of the genre, regardless of whether we take plausibility as a function of realism, as Watt has, or whether we take it to arise from fictionalization, as Molesworth proposes.[45]

Jane Austen recognizes this problem with the Richardsonian approach to action, devoting all her plots to the errors of first impressions and the necessity of reassessments through reenactment, rather than through better prospection. The importance of second takes in Austen has been amply and productively explored. Deidre Lynch aligns the "rereading" of characters that Austen's novels demand with Romantic practices of reading that define "literary experience" as a twofold recognition: that meaning of a character is dispersed temporally along narrative and that it "can never be definitively deciphered."[46] And Soni argues that rereading in Austen delivers a recognition that our hypotheses and the evidence to support these depend on our position in narrative time, which entails that judgment cannot be prescribed by either schemas or facts but can only be an activity that compares these as they variously cohere in any point in time.[47] Soni's reading of Austen clearly positions her within the tradition of the durational ethic that I have examined here. But I want to conclude this chapter with a brief reading of Austen's double takes in *Sense and Sensibility* (1811) in order to highlight the way in which Austen modifies this ethic, substituting the pure mental activity of virtuous projection for a more comprehensively sensible—mental and embodied—experience. In *Sense and Sensibility* all the main characters get a chance to relive their plot lines as though they are re-reading them. This novel values the second round precisely because it makes a comprehensive case for the necessary and unavoidable temporal recursive process of getting things right. In line with Hutcheson's and Richardson's durational ethic, in Austen getting things right pans out less as an epistemological certainty than as a lived experience. But if Hutcheson and Richardson imagine that we could stretch consciousness to the magnitude of action, then *Sense and Sensibility* programmatically rewrites forecasting as repeating plot lines—grafting the durational ethic not only on characters' minds but also on their bodies as these act and reenact.

To a certain extent the plot conceit of the second take serves as a critique of the libertine moment. Marianne's instantaneous judgments, general rapidity—her "extraordinary dispatch" (42) in conversation and her "active imagination" (50) that "speedily form[s]" (50) stories—and emotional absorption in the present, all resonate with the libertine privileging of the moment, which she learns by the end of the novel to eschew.[48] In its stead she adopts a more sedate nature and, most important, an openness to second loves. And yet Elinor's patience, lengthy deliberations, reluctance to form quick conclusions, and commitment to achieving a distant overview before judging and acting do not reward her with a successful first take or save her from a broken heart. Regardless of her deliberation, Elinor also needs a second take in order to revive her hope in Edward after she has been disappointed. By virtue of their occupying later durations, second rounds enable the disclosure of what might be taken as a steady set of attributes—Edward's understated personality, which Elinor takes the time to "study" (17), or Willoughby's complete set of motives, which neither sister could fully discover in the course of a first encounter. But even more important, second rounds enable the disclosure of trajectories of action already set in motion—Edward's prior commitment to Lucy Steele, Willoughby's rakish elopement with Eliza Williams. Both sisters—the impulsive and the deliberate—do not take such actions-already-in-motion into consideration in their first takes, and their second takes depend not simply on the disclosure of these trajectories of action, but also crucially on their resolution. Lucy must call off her engagement with Edward and thus free him to propose to Elinor; Willoughby's deceptive and harmful action must cost him his inheritance, thus dissuading him from further seducing Marianne and leaving her free for Brandon. At stake, then, is not simply coming to know things, but also experiencing the collisions of actions with trajectories of others' actions. But for Austen the painful results of such collisions do not mean that her heroines should have done more prospection; rather, it means that they should be prepared to relive their plot lines.

In Austen's account, first takes are the faulty, necessary prerequisites for second takes. Errors are not digressions, but are necessary for the forward movement of action. These errors and the narrative duration they take are necessary, but not as Hutcheson's and Richardson's virtuous yet impossible capacity for prospection on the move, but rather as a more plausible capacity for learning on the move. Austen thus recoups Richardson's re-reading for the plot into the diegesis and into action, making it a more easily apparent principle of plot. And she rewrites Richardson's thinking about action as a more thoroughly sensible experience. Her characters comprehend the ethical significance of linear and forward-moving trajectories of action not only, or even primarily, through their minds, but also through their bodies as these occupy the durations of their motions. In this way, *Sense and Sensibility* relies on and promotes the sensibility chronotope even more comprehensively than *Clarissa*.

## *Chapter 3*

# Sympathetic Moments and Rhythmic Narration

## Sterne, Early Musicology, and the Elocutionists

In a letter probably written in June 1761, Laurence Sterne depicts a passionate attachment characterized by a peculiar duration. Here is the letter's first paragraph:

> Of the two bad cassocs, fair Lady which I am worth in the world, I would this moment freely give the better of 'em to find out by what irresistible force of magic it is, that I am influenced to write a Letter to you upon so short an Acquaintance—*short,* did I say—I unsay it again: I have had the happiness to be acquainted with M$^{rs}$ Vesey almost time immemorial—surely the most penetrating of her sex need not be told that intercourses of this kind are not to be dated by hours, days or months, but by the slow or rapid progress of our intimacies which can be measured only by the degrees of penetration by w$^{ch}$ we discover Characters at first sight, or by the openness and frankness of heart w$^{ch}$ lets the by-stander into it, without the pains of reflection . . . the fair M$^{rs}$ Vesey—her character is to be read at once; I saw it before I had walk'd ten paces besides <you> her.—I believe in my Conscience, dear Lady, that you have absolutely no inside at all.— (137–38)[1]

An acquaintance that cannot be measured by "hours, days or months" but only by pace of feeling and mutual understanding, and a duration that is at once instantaneous and spans time immemorial—Sterne is indeed representing a moment that seems much more magical than probable. And yet as extraordinarily as Sterne describes this moment and its momentum—by which I mean its presence as well as is its force, rather than the kind of linear unfolding that we tracked in the previous chapter—he also flags it as delivering the feeling of utter certainty and perfect communicability. Mrs. Vesey ought already to know exactly what Sterne is talking about—"surely the most penetrating of her sex need not be told"—while Sterne already fully knows all that Mrs. Vesey thinks and feels—"you have absolutely no inside at all." Sterne characterizes his passionate moment as retiring the need for verbal explanation. And yet this peculiar sympathetic bond requires a communicative medium and, even more, a linguistic one. Not only is Sterne compelled to write a letter, but he also presents the moment as romance and as a transaction between characters. He begins his address by referring to Mrs. Vesey as the stock "fair Lady" and by conflating himself with Tristram Shandy. "Of the two bad cassocs" and so on is an almost verbatim—albeit unremarked—quotation of Tristram's aside on Toby's response to Dr. Slop's musical curse of Obadiah. And in the letter's last paragraph Sterne denotes his addressee by a letter rather than by her full proper name, and he refers to himself as his character: "My dearest M$^{rs}$ V. what business you had to come here from Ireland—or rather, what business have You to go back again—the deuce take you w$^{h}$ your musical and other powers—could nothing serve you but you must turn T. Shandys head, as if it was not turn'd enough already" (138). Sterne thus constructs his emotional bond with its peculiar duration as essentially discursive; it's an experience that seems to be wholly—even exclusively—available in language.[2] And yet, Sterne indicates an additional medium underwriting the moment of passion with its utter communicability; if Mrs. Vesey is instantaneously known and beloved by Sterne, this is because he perceives her qualities as a connoisseur perceives music:

> That you are graceful, & elegant & most desirable & c & c. every common beholder, who only stares at You as a dutch Boore does at the Queen of Sheba in a puppit Show can readily find out; But that You are sensible, and gentle and tender—& from end to the other of you full of the sweetest tones & modulations, requires a Connoisseur of more taste & feeling—in honest truth You are a System

> of harmonic Vibrations—You are the sweetest and best tuned of all Instruments. (138)

Anyone can behold Vesey and perceive some of her qualities; but Sterne more than sees her. He hears her—a perceptual mode that underlines both Sterne's refined capacities and Vesey's identity. For if Sterne must be "a Connoisseur of more taste & feeling" to apprehend Vesey in this way, then Vesey must be—not merely be like—"a System of harmonic Vibrations" and the "best tuned of all Instruments."

This chapter explores such a conception of "this moment" as a peculiarly sympathetic duration constituted as both music and discourse. At stake is an approach to time that neither assesses trajectories, as in Francis Hutcheson's and Samuel Richardson's durational ethic, nor measures expanses, as in chronometric approaches. Rather, it conceptualizes duration in terms of frequencies and rhythms, vibrations and pulses to conceive the very material constitution of what David Hume calls, as we've seen in Chapter 1, the sensible moment. This sensible moment is experiential and particular, but it is also fundamentally shared by virtue of its rhythmic pattern and communicability through sound, a sound that can be as much linguistic as purely musical. And while such moments may sometimes seem indistinguishable from erotic transactions—like Lovelace's sexualized sympathy and as Sterne never fails to remind us in his letters and in his novels—many writers insist on their peculiar conduciveness to properly sociable fellow-feeling.

Even Adam Smith, in his influential theatrical and mostly spatial model of fellow-feeling, occasionally uses a musical temporalized vocabulary. Thus he explains the desire for "a more complete sympathy" felt by a person in distress as a longing that "the emotions of their [other people's] hearts, in every respect, beat time to his own." The sufferer can actualize this longing, at least partially, as Smith tells us, "by lowering his passion to that pitch in which the spectators are capable of going along with him. He must flatten . . . the sharpness of its natural tone, in order to reduce it to harmony and concord with the emotions of those who are about him" (22).[3] In another example Smith recommends a deliberate adjustment of pulse, here placing the burden of pacing on spectators: "Our heart must adopt the principles of the agent, and go along with all the affections which influenced his conduct, before it can entirely sympathize with, and beat time to, the gratitude of the person who has been benefitted by his actions" (73). Such moments in *The Theory of Moral*

*Sentiments* (1759) highlight bodily organs, rather than cognitive capacities—hearts beating synchronously and pitches concerting harmoniously, instead of imagining ourselves on the rack when we see someone else there. These moments also privilege people doing something together, instead of each individual acting a different role across the spectator-spectacle divide—"going along" versus observer and observed. And though a spatial vocabulary remains in play through the musical metaphors of lowering and flattening, the process Smith describes clearly includes the temporal underpinnings of frequencies and pace in the form of beating time and tonal adjustment.

It is hard to tell whether Smith, like Sterne insisting that Vesey *is* a system of vibrations and a tuned instrument, means that hearts beating in tandem literally constitute our sympathies, and that our emotions physically come to correspond by tuning their pitch. Smith's overall emphasis on mental images may incline us towards a figural interpretation of his proposition. And yet we might also recall that the physiological and psychological theories underwriting sensibility conceived of perception as a process of vibrations. G. J. Barker-Benfield points to Isaac Newton as the source of this model, explaining that "while he may have derived his notion of the 'Sensorium' (the termination of all the nerves in the brain) from Willis, Newton differed from Willis's idea of the hollow nerve through which animal spirits flowed from brain to organs. Instead, Newton had argued from 1675 that the nerve was solid and transmitted sense impressions by vibrations" (4). By the logic of perception as a vibratory mechanism, sensation and music share the same form—that of dynamic matter organized by recursive pulses. Relying on this model, George Cheyne explained that "the Intelligent Principle, or *Soul,* resides somewhere in the Brain, where all the Nerves, or instruments of sensation terminate, like a *Musician* in a finely fram'd and well tim'd Organ—Case; these Nerves are like *Keys* which, being struck on or touch'd, convey the Sound and Harmony to this Sentient Principle, or *Musician*" (7–8).[4] If Cheyne begins with the "Soul" as analogous to a "Musician," he concludes with "Sentient Principle" as fully interchangeable with "Musician"; the simile turns into identity, indicating not simply the slipperiness of Cheyne's thinking but, rather, the extent to which this vibratory model takes musical experience as paradigmatic of human experience more generally.

If late seventeenth- and early eighteenth-century philosophers and anatomists sometimes analogize human experience to music and sometimes fully identify the two, then early musicologists bolster the case for full identification.

As my discussion in the first section of this chapter demonstrates, early musicology attaches our primary sense of duration to physiological pulses, aligning it with a natural apprehension of rhythm. Such a definition of duration significantly revises John Locke's, making temporality at once embodied and shared. Rather than arising from the succession of ideas passing in individual minds, our sense of time relies, according to early musicologists, on isochronic motions that naturally animate the human body. Such an understanding of musical time is extended into language by midcentury treatises on elocution. These treatises, I show in the second section of this chapter, make an explicit case for the way in which sonorous and, specifically, rhythmical speech underwrites fellow-feeling. Like the musicological model, elocutionist discussions stress regular beats as the most essential element of rhythm. They also suggest that while clocks and calendars can effectively measure length and offer abstract notations of time, rhythm assesses pace and pattern and is never complete without a pulsating body to perform and experience such renderings. With this experiential yet formal conception of rhythm and duration in mind, I return to Sterne's renderings of sympathetic moments in the last two sections of this chapter. If in the letter to Vesey, sensibility's musico-linguistic sympathetic moments appear as shorthand, in *Tristram Shandy* (1759) these are fully developed through pulsating prose, philosophical meditations on time, and experiments in narrative duration. Sterne highlights the extent to which the prose of novels can convey the rhythms of language and thus represent moments of sympathetic immersion between characters and mediate such durations for readers.

## Musical Duration and the Inner Time-Sense

In an interview with Glenn Gould on the occasion of Gould's second recording of the *Goldberg Variations*, Tim Page observes that when first hearing the new performance, he was struck by its exceptionally slow pace. Page then confesses to timing the recording, leading him to discover that the 1981 performance is only thirteen minutes longer than the 1955 one. This might count for much in music, if it weren't for the many repeats added in the second version; subtract repeats from total duration, and the two recordings are pretty much the same length. Nonetheless, the 1981 performance seems significantly slower. "Frankly I can't figure it out," Page concludes his question. Gould's response, which includes precise calculations and demonstrative singing, comes down to this: the second recording implements a technique of rhythmic continuity, with one constant rhythmic reference point underwriting the entire performance. The theme of the *Goldberg Variations*—indeed, played significantly slower in 1981—sets a pulse rate for the rest of the recording, retained in subsequent variations by divisions and multiplications that make subsidiary beats. Throughout these varying tempi the feel of the basic pulse-rate persists, fostering an experience of dilation, even when the actual playing pace is quite fast. But in conclusion, Gould emphasizes: "When one describes a process this way it sounds just so relentlessly clinical, so ruthlessly sterile and anti-musical. . . . A technique, the idea of rhythmic continuity, that's really only useful if everybody does feel it in their bones, to use your words; experiences it subliminally, in other words."[5]

Such a conception of duration keyed not to clock but to rhythmic pulse, and at once underwritten by formal precision and embodied experience, might be traced to early eighteenth-century musicology. Roger North in his *The Musicall Grammarian* (1728) highlights the importance of equable beats in constituting musical time and indicates physiological sources for such rhythmic pulses. He explains: "For equallity takes place in our most thoughtless actions, as when wee walk, run, strike, act with our fingers, or swing an arm, all run in equallity" (126–27); "This simple equallity is the root, out of which all the varietys of musciall time spring" (127), he adds and continues: "All expression of melody or tune is the result of distinguishable sonorus pulsations in a series continued, and all leaning upon some gross equallity of time. As when at distance wee hear a company dancing litle is perceived but equall timed stepps, and being present wee can scarce forbear pulsing the like, and the effect of art is onely diversifying these measures by affected subdevisions" (127–28).[6] Here as elsewhere in his work, North argues that rhythmic duration is somehow lodged in embodied experience—whether in our ordinary bodily actions (walking, swinging an arm) that rely on patterns of isochronic pulsations, or in our unstudied responses to temporal art (dance, music), which recruit us instinctively as we "scarce forbear pulsing the like." He thus suggests that rhythms of external dynamic objects reflect for us an inner rhythmic time-sense that governs our own constitution. He writes: "Nothing unequall timed is pleasing, as if equallity had planted its capitall residence in our vitall facultys" (127). From this perspective, isochronic pulsations residing in our own "vitall facultys" make musical rhythm less a metaphor for our temporal experience than an external index of the very principles that structure our sense of duration.

Jamie Croy Kassler has traced North's conceptions of time to Thomas Hobbes's and Robert Hooke's mechanistic materialisms. Hobbes conceives of bodies as elastic, vibrating substances, and Hooke claims that we perceive such vibrations by responding with similar motion in our peripheral sensory equipment such as the eardrum and the pupils of the eyes that, then, continues to travel as tremulous motion along the nervous system to the brain, which itself also vibrates. North similarly argues that our perception comprises a series of vibratory motions—from the outer perceptual apparatus, through the pulsing nervous system, to the brain—all responding to pulsing motion of air excited by vibrations of external bodies that appeal to our senses.[7] Moreover, both Hooke and North distinguish between vibrations that we perceive as such, and those frequencies of infinitesimal make-up that we register only unconsciously. Hooke calls the threshold of consciously perceptible frequencies "one human Moment," and North calls the threshold of consciously *im*perceptible frequencies "notable pulses."[8] Thus for Hooke and North, the vibratory constitution of matter and of perception underwrites our sense of duration, defining a moment as a recognizable pitch. And the conception of a moment as such—pulsating form appearing as sound—makes it possible to understand music as an external analogue of our primary temporal experience. By this logic, musical sounds correspond to durations whose rhythmic inner constitution we do not consciously calculate, whereas musical time corresponds to durations whose rhythms we can abstract even as such abstractions come to us intuitively.

This account offers a different approach to the primary experience of duration than the one Locke presents in *An Essay Concerning Human Understanding* (1689). As we've seen in Chapter 1, Locke distinguishes between an inner time-sense and a community's conventions for representing history and locating events in time; a person's primary "idea of duration" arises from his own successive reflections and, as such, is essentially individual and psychological, while a community's conventions for temporal representation rely on a principle of periodicity that is conventional and abstract. We've also seen that Locke's commitment to the perceptual origins of ideas can be brought to bear on his account of duration, such that it troubles his commitment to abstract measures and privileges a sensible durational experience. Yet Locke never argues that periodicity can be derived from physiological tendencies, nor that our sense of the duration of an instant or a moment might arise from vibratory motions in our nervous system. By stressing a subliminal tendency for equable motion, North replaces Locke's succession of ideas with a rhythmic beat; by lodging such pulsations in bodily movements, he makes the primary experience of duration physiological rather than psychological; and by identifying such embodied rhythms as natural, he suggests that inner temporal experience is, to begin with, shared—spanning a limited range common to human bodies.

For North, then, the primary experience of duration is subconscious—we sense time involuntarily through instinctive bodily pulses. And he continues to suggest that the making conscious of this subconscious experience requires practice and technique. Thus while North insists on the immediate visceral effects of music, he also emphasizes that keeping musical time is by no means an easy skill: "Of those difficultys none are found greater then to attain a true knowledg and setled habit of keeping musicall time" (126). And he continues: "Altho wee are so naturally addicted to it, yet in the artfull practise, no part is so hard to acquire with sufficient exactness, as that which belongs to time" (127). The instinctiveness of rhythmic pulsations makes their conscious rendering as art all the more demanding. And thus, much like Hume's suggestions about the training of our intuitive temporal measures, North also recognizes that precision and mastery in musical timekeeping requires practice.

Many eighteenth- and nineteenth-century musicologists recommend practicing with the support of tools such as pendulums, stopwatches, and metronomes to stabilize pace by a mechanical beat. From Galileo Galilei's experiments with pendulums, which he began in 1602, to Johann Nepomuk Maelzel's patenting of the metronome in 1815, we find in musical treatises an increasing competition between pulsating timekeeping devices and heightened attention to natural bodily motion, with most writers recommending some combination of the two. Perhaps the most startling insistence on such combination is Edward Hodges's endorsement of "an electrical machine, which was to give the pupil a moderate shock, without noise, at the commencement of each bar."[9] Whether satirical or serious, this contrivance merges mechanical and embodied timekeeping; with moderate electric shocks, students of music in a mechanical age may be reminded of their primary embodied sense of duration. If chronometric and chronological timekeeping rely on abstract conventions (which, as we've seen in Locke's discussion, do not require and in fact make superfluous, personal experience), then musical synchronization serves as a reminder, or arousal, of our embodied primary sense of duration. But as we will see, a less violent and more commonly available reminder of our inner time-sense than the one proposed by Hodges was imagined by English elocutionists. These advocates of oral language and the standardization of English pronunciation made the case for the physiological sources and impact of speech and its rhythms.

## Linguistic Rhythm and Sympathy

In multiple lecture series across Britain and numerous publications on correct pronunciation and effective delivery, English elocutionists redefined the role of oral language for an increasingly print-dominated culture. One important dimension of their project was to standardize speech, an effort that many scholars have understood as the imposition of pre-set forms on individual speakers.[10] But the elocutionists also used sophisticated means to trace the ways in which language arises from primary embodied experience, rather than simply imposing social norms on bodies. And, equally important, they made a strong case for speech's capacity to generate a sympathy anchored not in spectacle and convention but, rather, in common aural and physiological patterns. Much of this more subtle and complex tack on the relation between bodies, language, and sociability can be found in elocutionist discussions of rhythm.[11] Most elocutionists consider rhythm as a primary dimension of linguistic efficacy. Thomas Sheridan suggests that rhythm distinguishes language from mere babble, and John Herries recommends beating time to speech as the first and foremost exercise in cultivating good listening and speaking habits.[12] But it is Joshua Steele who offers the most extensive and sophisticated analysis of linguistic rhythm in his *An Essay Towards Establishing the Melody and Measure of Speech* (1775), where he highlights physiological dimensions and indicates how such embodied duration can be communicated not only in presence but also in print.

Steele explains that musical time is measured not by expanse, but "by *pulsation*, quicker or slower. The *pulsation* of any one sort of *time* must continue as uniform as the swinging of a pendulum of a given length; but the intervals between the *pulses* of the pendulum may be filled variously" (69).[13]

And as in the musicology he draws on, Steele's preferred isochoronic pulses that tick from within the bodies of speakers:

> The beating of our *pulse*, which we must feel whenever we are silent and inactive, prones us to *rhythmical divisions* even in the series of our thoughts; as soon as we begin to move, our *steps* succeed in the government of *rhythmical pulsation*, and the *measure* may then be at our *option* faster or slower; for while we were silent and motionless, the *measure* of our thoughts must have been regulated by the *cadences* of our *pulse*, which is an involuntary motion.
>
> Every single *step*, or every pace, may mark a *cadence*, the putting down the foot being *heavy*, and the lifting it up being *light*. . . .
>
> If our *pulse* is to govern the time or length of a *cadence*, the *thesis* and *arsis* must keep pace and coincide with the *systole* and *diastole* of the heart. (118)

Our pulse, then, gives rhythm to our thoughts—a point that enables Steele to reconcile the varying Lockean and musical accounts of duration. If "time is measured by *pulsation*" and "our *pulse* . . . prones us to *rhythmical divisions* even in the series of our thoughts," then the inner time-sense supported by the succession of ideas registers the rhythmic divisions of our pulse as it punctuates our thoughts.

David Abercrombie notes that modern physiological linguistic research supports "the view, implicit in Steele's theories, that rhythm is not something to be found in the *sounds* of speech, but rather in the muscular movements of the speaker and the listener's sympathetic knowledge of those movements."[14] Abercrombie is right to emphasize that for Steele, as for other elocutionists, linguistic rhythm crucially attaches to bodies and cannot be reduced to prosody. But also worth noticing is that Steele rigorously works out how embodied beats *interact* with formal sound patterns of language to produce a sense of rhythm that can be communicated through speech and through writing and, thus, can be counted on as shared by all who use any given language. For Steele, rhythm is at once embodied and linguistic, and such double grounding makes rhythmic language especially effective at conveying sympathy. But this also entails that for the elocutionists, the feeling shared in language runs along the borders of national languages.[15] Steele argues that the way in which intervals between pulses are filled is as important in the production of rhythm as the pulses themselves; in language these intervals are filled with patterns of sound and silence determined by the emphases appropriate to words in each language—a pattern that Steele calls "poize." The word "impossible," he explains, gains its rhythm in English from beginning with a syllable of light emphasis, proceeding to one with a heavy emphasis, and then another two with light emphases. Beginning with a light, it must always start off beat and "cannot be twice repeated without leaving the *quantity* of half a *cadence* in silence" (119). (Steele defines "cadence" as "the space of time between each pulsation and the next succeeding pulsation" [116] and explains that "cadences always begin with *thesis*, or the *heavy* syllable, and end with *arsis*, or the *light*" [115].) He contrasts this rhythm of the English "impossible" with French, in which, "impossible" begins with a heavy syllable, proceeds to a light, another heavy and another light, thus filling a full two cadences with sound and generating a different rhythmic pattern—heavy, light, heavy, light, versus the English light, heavy, light, light (120). We might pronounce the heavy syllables for a longer duration than the light ones (and sometimes Steele implies that we do), but such comparison does not constitute the core of rhythmicality in his understanding:

> In the *rhythmus* of language, all polysyllables are affected to their *poize* of *heavy* and *light* so positively (and the *poize* determines the *cadence*), that nothing remains in doubt except the difference between the slowest and the fastest speaker. However, that doubt is of no consequence in this argument, since every speaker, if he preserves the proportions demanded by the natural *quantity* and *poize* of the words, must adopt that measure of quickness that the *poize* of the words points out. (119)

Speakers' speeds may vary, and thus the lengths of syllables may also vary; remaining constant, though, is the proportionate relations among syllables relative to pulse or step, and the expanses of these proportionate relations are severely limited by patterns of heavy and light emphases appropriate to the standard pronunciation of words in each language. Individual variations are, thus, bounded by a range that secures the recognizability of a language's distinctive rhythm.

Furthermore, Steele suggests that knowledge of this range comes naturally to native speakers. He declares "our sense of *rhythmus* to be much more *instinctive* than *rational*" (78), repeatedly remarking that "we have an instinctive sense of *rhythmus*, as connected with, and governing, all sounds and all motions" (67). While nonnative speakers might fail to perceive the beats and emphases of any given word, natives "find no difficulty to distinguish and evaluate notes so minute as twenty-four or thirty-two to a second of time" (156) without ever programmatically learning how.

Although most elocutionists have a more didactic agenda than Steele, they tend to agree about the instinctive constitution of rhythm and the sensible qualities of durational sympathy. Even the most prescriptive of the elocutionists avoid specifying guidelines when it comes to language's pulsating patterns. Herries, for example, prescribes everything from proper body position, to adequate breathing, to correct tone and pitch. However, when it comes to rhythm, he maintains: "Pronounce any one line from Milton, and the EAR will determine whether or not the accent and quantity always coincide. Very seldom they do" (123).[16] And he continues: "I appeal to the ear as the best standard to determine which syllables are long and which are short in pronunciation" (128). Sheridan similarly remarks of such words as "concordance," "refractory," and "accented": "the accenting of these being doubtful, every man is at liberty to choose which he likes best; and in giving the preference, the ear beyond all doubt ought to be consulted, as to that which forms the most agreeable sound, rather than an absurd, pedantic rule, attempted to be laid down . . . which has no foundation in the genius of our tongue."[17] Sheridan begins by privileging individual expression—"every man is at liberty to choose which he likes best"—but quickly veers to prioritizing systemic determinations: any preference ought to be anchored "in the genius of our tongue." This suggests that rhythm defies prescription not because its determination ought to be left to individual speakers, but because it is already decided by concrete experiential—Steele's "instinctive"—preferences, which are typical. For the ear in Sheridan's discussion, as well as in Herries', stands neither metonymically for the deliberating individual nor metaphorically for the person's mind or taste. Rather, both Herries and Sheridan in these examples refer to the organ of perception itself as it reacts in response to material stimuli. Hence Herries' "I" can "appeal to the ear," rather than itself be the ear.[18]

Herries also criticizes those trying—impossibly, by his account—to set modern English to classical rules: "Others have attempted to reduce the quantity of our syllables to the standard of the Antients. But how can we propose a standard of which we are entirely ignorant? If we had *heard* Homer recite his verses, or Cicero pronounce his orations, we then might judge with certainty concerning the length of the Greek and Latin syllables. But as we have never had that opportunity, we must give up all pretensions to this species of critical knowledge" (123–24, emphasis added). Herries argues not simply that classical rules are irrelevant to the rhythms of modern languages, but that these are altogether unavailable to modern English speakers because they do not have direct perceptions of speech in either classical Greek or Latin. When it comes to rhythm, even the most diligent rational surmises remain unreliable; without the presence of his ear at oral recitations by native speakers, even the most erudite modern scholar can only remain ignorant of the ancient standard. But this doesn't mean that linguistic rhythm eludes form as such. Greek and Latin have rhythmic standards, which merely elude the "pretensions to . . . critical knowledge" in the absence of direct sensory experience.

I am arguing that the difficulty of prescribing rhythm, according to these writers, arises not from demands of standardization but, rather, from the attempt to formulate standards as abstract rules. Adequate durations and accents in speech elude principles that can be articulated independently from their embodied experience; rhythm, Sheridan and Herries indicate, cannot be numerically represented, but must be sensed and felt. Yet this doesn't make rhythm individual or irregular; rather, it casts it as an embodied version of systematicity. And in this way the elocutionists bypass the problem that, as we saw in Chapter 1, Locke runs into when he wants to privilege primary durational experience while he also insists that shared temporal measures are abstractions that make experience superfluous. For the elocutionists, as for early musicologists, rhythm neither eludes standardization nor becomes alienated by such shared measures; instead, it exemplifies an experiential fit between individual habits and collective standards that can be neither generated nor communicated without the physiological pulses and predisposed ears of particular speech communities.

Herries declares that "there is such a mutual sympathy between the ears of the audience, and the organs of the speaker, that whatever seems to give pain to the latter, will produce the same sensation in the former" (121–22). In such an account, at stake in fellow-feeling is less the sharing of semantic content than the synchronizing of embodied experience, something that the sonorous dimensions of language achieve especially powerfully. And, in turn, such an understanding of sympathy entails that by standardizing pronunciation a nation enlarges the numbers of its own members who have similar impressions of linguistic sonority and who will, thus, be able to partake in fellow-feeling. Different from understandings promoted by recent criticism of eighteenth-century national liberal projects, for elocutionists such as Steele, Sheridan, and Herries, standard pronunciation and performed gestures do not eradicate authenticity, nor do they privilege an enforced surface commonality at the expense of an interior, repressed individuality. Rather, such standardization and performativity extend the reaches, in number of participants as well as over distances of space and time, of sympathetic communication.[19]

For the purpose of such extension, Steele devises ways for rendering rhythm in writing, which include not only the use of innovative typeface but also the insistence on voicing and dancing the print that we read. First, Steele recommends that writers specify pace for voicing their written words, just as composers indicate tempo of performance for their music: "The writer may follow his judgment or fancy in fixing the degree of velocity, by marking it for two steps to a second of time, or one step to a second, or more, or less; or he may leave a greater latitude, by marking it *slow walking time, moderate walking time, quick walking time, or running time*" (121). Steele continues to develop an elaborate system of notation to represent all the subtleties of language's musical dimensions. He proposes one set of signs to indicate accents and pauses and to be placed above words, another set of signs to indicate heavy and light syllables and to be placed below words, and bars marking the beat to be placed among and within the words. A written phrase, then, would look like Figure 1 (122).[20]

Andrew Elfenbein points out that such remarkable use of typeface, which can also be found in Sheridan and Walker, indicates the extent to which print overtakes the elocutionists' privileging of voice. "Elocution," Elfenbein

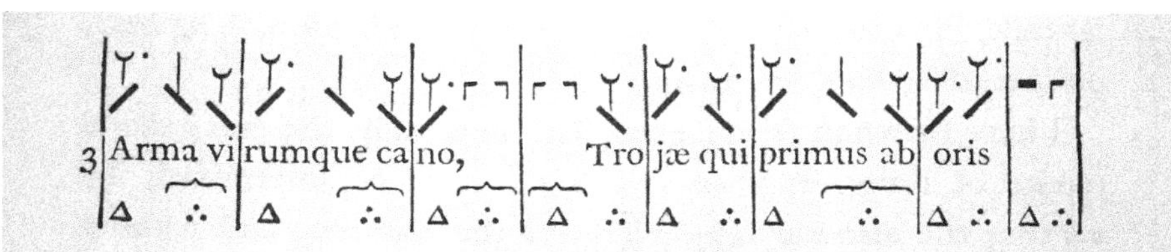


Figure 1. Joshua Steele's system of notation to represent the musical dimension of language. From Steele, *An Essay Towards Establishing the Melody and Measure of Speech* (1775).

argues, "became a technology of transmission, an interface for translating one medium (print) into another (voice) for the benefit of an audience."[21] If earlier scholars frame the elocutionary project within a competition between print and orality, writing and speech, then, as Elfenbein contends, by the mid-eighteenth century the medium of print could no longer be conceived as a foe to beat, but only as one to join. This is an important observation, but it is also worth noting that for Steele such translation of print into voice goes two ways: to his sophisticated notation, he adds the insistence that elaborate marks be translated back into performance. Steele really means readers should pace and voice—quickly, moderately, or slowly—what they read:

> Then let the reader walk and pronounce, putting down (suppose) the right foot to *ar*, lifting up the other with *ma vi*, down again on *rum*, right foot up to *que ca*, down to *no*, the left up under the *pauses* ┌┐, and down under the *pauses* ┌┐, the right up to *Tro*, down to *jæ*, left up to *qui*, down to *pri*, right up to *mus ab*, down to *o*, left up to *ris*, left down and the right up to the pauses of = minim and ┌ crochet. (122)

Had Steele only invented a typographical system (ingenuous as it may be) and suggested that authors ought to recommend a range of reading paces, his innovations would have merely enabled readers to imagine rhythm more precisely, rather than actually to experience the rhythmic interaction of pulse or step with poize. Sophisticated notation constitutes an improved medium for representing rhythm without creating a concrete embodied experience through writing. Yet Steele does not give up on such experience. He aims at the literal synchronization of pulses, even if such synchronization occurs only at a distance (both the spatial and temporal distances presumed by writing). And we shouldn't be surprised that Steele's example for fully performed writing is not in English (he uses a quotation from the Latin *Aeneid*). Since rhythm is instinctive in Steele's account, an Englishman will need such meticulous instruction only in a foreign language. When an Englishman reads English print, he simply needs to get up and voice and dance the text, observing the author's pace recommendations. Patterns of pulse and poize in our native tongue are likely to come out right without precise instruction as to how movement and words ought to be coordinated, provided that we actually perform the physical movement represented in the notation.

To observe the importance of voicing and dancing our reading is also to recognize that the elocutionists' understanding of sympathetic language differs from more familiar accounts of sympathetic spectatorship which rely on both fiction and dissociation. As David Marshall points out, in the theatrical model we imagine others' feelings as we behold them, rather than get close and immerse ourselves in their experience.[22] Steele's model, by contrast, does not posit performers on one side of the communicative circuit and beholders on the other; rather, it asks for active performance of all who are involved. Unlike beholders, readers must dance what they see, and such performance ensures that readers' bodies synchronize with the rhythms intended by writers. Yet it is worth emphasizing again that such successful sympathetic communication relies on the restriction of messages to the forms made possible by language. In the example from *The Theory of Moral Sentiments* with which I began this chapter, sympathetic communication depends on bringing one's pulse rate back within the range of ordinary speakers; for the elocutionists, sympathetic communication never transmits raw prelinguistic content, but only the embodied experience of the poize and pulse of language and its users. The "impossible" that you and I can share is not a condition or feeling that stands as a referent independent of our humanity and of the sound pattern used to signify it; instead, it is the referent as it is already both embodied and formalized by its specific rhythm in any given language.

## Sterne's Pulsating Prose

William Wordsworth famously claims for his poetry the capacity to communicate emotion in verse that most resembles ordinary speech. And though he insists that both prose and verse "speak by and to the same organs; the bodies in which both of them are clothed may be said to be of the same substance, their affections are kindred, and almost identical, not necessarily differing even in degree" (69), he also observes that "the only strict antithesis to Prose is Metre" (69) and points out that meter is "something which will greatly contribute to impart passion to the words" (82).[23] Wordsworth's account of language's capacity to convey sympathetic moments through meter has been extraordinarily influential, making it somewhat difficult for us to recognize other forms of rhythmical language and the ways these have been mobilized in prose.[24] Following the importation of a beat-based conceptualization of musical time into elocutionist discussions of language enables us to identify an approach to linguistic rhythm that does not rely on meter. And close attention to Sterne's *Tristram Shandy* can demonstrate how Steele's expansion of musical time to a theorization of linguistic rhythm helps us better recognize a pulsating prose and its effects on narrative duration.

My analysis of *Tristram Shandy* doesn't presume that Steele directly inspired Sterne. Sterne published his fiction between 1759 and 1768, before Steele published his *Essay*; thus, if there are any lines of direct influence, they would have flowed in the opposite direction. Indeed, there are several moments in Steele's *Essay* that read like exemplars of Sternean humor—when readers are directed to dance and voice what they read, or when Steele remarks on riding habits that make us as prone to sympathize with horses as with our fellows humans— "our habit of riding makes us almost as familiar with the measures beaten by the paces of horses as if they were our own" (126), he writes. But in any case, rather than claiming direct influence in whatever direction, my point is to show that Sterne and the elocutionists share a complex of related interests in duration, rhythm, sympathy, and language. Or, in other words, that they all participate in developing this dimension of the sensibility chronotope.

In *Tristram Shandy*, Sterne polemically engages Locke on his theory of language and his theory of duration. In line with early musicology's accounts of the inner time-sense and with the elocutionists' conceptions of linguistic rhythm, Sterne carefully displaces Locke's contention that the succession of ideas is the source of our primary experience of duration. But before we can fully understand Sterne's account of duration, we need to look at how he challenges Locke's theory of language.[25] In the *Essay* Locke argues that when words fail to attach adequately to ideas, "they are nothing but so much insignificant Noise" (408).[26] "It is a perverting the use of Words, and brings unavoidable Obscurity and Confusion into their Signification, whenever we make them stand for anything, but those *Ideas* we have in our own Minds" (407), he maintains. For him, language is "the great Instrument, and common Tye of Society" (402), made of "external sensible Signs, whereby those invisible *Ideas* . . . might be made known to others" (405). And yet throughout *Tristram Shandy*, and contrary to Locke's position, Sterne equivocates the relation between words and ideas and makes the case for words' serviceability to the communication not of knowledge but of feelings. Early on Tristram declares Locke's *Essay* "a history-book . . . of what passes in a man's own mind" (1.98).[27]

This is an especially literal application of Locke's logic, attaching his words to his ideas so insistently that it makes philosophy comparable to a novel. For if Locke's *Essay* is "a history-book . . . of what passes in a man's own mind," then it is of the same class as *The Life and Opinions of Tristram Shandy*. This proposition is both an honest description of Locke's book and a joke about it—and as such, it cannot itself be exemplary of Locke's principles, which privilege unambiguous expression. A little later in the same chapter, Tristram explains to a critic who questions Toby's character that "'Twas not by ideas,——by heaven! his life was put in jeopardy by words" (1.101). Here also, it is not clear whether Tristram means that the critic mistakes Toby's words for his ideas—making Toby's words confused but his ideas clear, thus detaching one from the other—or whether Tristram means that Toby's words confuse his ideas—making the two distressingly attached to one another. But both possibilities turn the attachment of words and ideas from an obvious value into a problem—an improbability or a dangerous possibility. All this, of course, confirms the centrality of doubleness to *Tristram Shandy*.[28] But to say this only serves as more evidence for this novel's skeptical attitude toward the relation between words and ideas: a book may run for many sentences, be an instant best seller, and continue a favorite all over the world for more than two centuries precisely by ambiguating the ideas of its author. Yet in the same chapter that directly problematizes Locke's dictates regarding the correspondence of words and ideas, Sterne points to the attachment of words and feelings as a more adequate pairing. To the critic who claims that Toby's confusions are an improbability, Tristram retorts that instead of passing judgment he ought to "drop a tear of pity" (1.100). All this language, Tristram suggests, does not communicate knowledge, but rather elicits sympathy—moves readers' bodies in ways that we associate with emotional expression.

Throughout *Tristram Shandy* words serve to forge less a meeting of minds than a meeting of feelings, and often such sympathy is rendered as shared moments of especially pronounced rhythmic qualities. Moreover, Sterne offers a thoroughgoing account of the inner time-sense as determined by such sympathetic linguistic sonority. Consider Walter and Toby's conversation while Tristram is being born. Walter confesses a hostility to clocks based on their eclipsing of the inner time sense: "In our computations of *time*, we are so used to minutes, hours, weeks, and months,——and of clocks (I wish there was not a clock in the kingdom) to measure out their several portions to us, and to those who belong to us,——that 'twill be well, if in time to come, the *succession of our ideas* be of any use or service to us at all" (1.224), he laments. But the irony of Walter's anxiety rests in his strong attachment to ideas and, thus, in his missing of the way in which the chapter, far from presenting the primary experience of duration as an endangered species—one that requires the kind of violent reminders of Hodges's electrical shocks—confirms its persistence through conversation.

The chapter begins with Walter's dismay at an ill correspondence between clock time and his own durational experience: "It is two hours, and ten minutes,—and no more,——cried my father, looking at his watch, since Dr. *Slop* and *Obadiah* arrived,——and I know not how it happens, brother *Toby*,——but to my imagination it seems almost an age" (1.222). Tristram continues:

> Though my father said, "*he know not how it happen'd,*"——yet he knew very well, how it happen'd;——and at the instant he spoke it, was pre-determined in his mind, to give my uncle *Toby* a clear account of the matter by a metaphysical dissertation upon the subject of *duration and its simple modes*, in order to shew my uncle *Toby*, by what mechanism and mensurations in the brain it came to pass, that the rapid succession of their ideas, and the eternal scampering of the discourse from one thing to another, since Dr. *Slop* had come into the room, had lengthened out so short a period, to so inconceivable an extent. (1.222)

While Walter sets out to lecture Toby on Locke's definition of the primary sense of duration, Tristram's narration makes an addition to the philosopher's analysis: duration not only arises from the "succession of their ideas," but also from the "scampering of the discourse." Sterne flags strings of linguistic sounds, and not just strings of thought, as crucial for understanding how temporal experience works. And throughout the chapter the brothers remain responsive to each other's words while being oblivious to these words' meanings. To Walter's "I know not how it happens," Toby replies, " 'Tis owing, entirely . . . to the succession of our ideas," thus frustrating Walter by speaking an explanation whose meaning Toby couldn't have possibly thought. And they continue:

> Do you understand the theory of that affair? replied my father.
>
> Not I, quoth my uncle.

> —But you have some ideas, said my father, of what you talk about.——
>
> No more than my horse, replied my uncle *Toby*.
>
> Gracious heaven! cried my father, looking upwards, and clasping his two hands together,——there is a worth in thy honest ignorance, brother *Toby*,—'twere almost a pity to exchange it for a knowledge. (1.223–24)

Toby has no idea of what he talks, and he easily admits it. Meanwhile Walter, although ostensibly committed to a close correspondence between words and ideas, actually continues to drive them apart; for as he declares it "a pity" to dispel Toby's "honest ignorance," he proceeds to lecture his brother about duration, offering a close paraphrase of Locke's *Essay*. Ideas continue to be displaced by words as Toby responds eagerly to Walter's saying "train" (1.225), but his answer only leads Walter away from "ideas": "Now, whether we observe it or no, continued my father, in every sound man's head, there is a regular succession of ideas of one sort or other, which follow each other in train just like——A train of artillery? said my uncle *Toby*.—A train of a fiddle stick!—quoth my father" (1.225). And as Toby's associative train turns out not simply to be different from Walter's, but more specifically to lead away from ideas, one cannot help but note the double meaning of "every sound man's head," whereby, contrary to Walter's intention, good judgment may rely not on coherent concepts but on resonating sonority. Walter's imperviousness to this kind of soundness finally gains the upper hand as he terminates the discussion having been convinced that none of his ideas were communicated in his discourse: "I have nothing more to say to you upon the subject" (1.225), he declares. Yet this statement also cannot escape doubleness; even if Walter ends the conversation because he wants to emphasize the importance of staying on message—the coherence of ideas that make for a discussion "upon the subject"—his statement also evokes its negation—the "eternal scampering of the discourse" that makes for plentiful conversation off topic. For in this chapter Sterne explains how two hours and ten minutes can seem an age by way of none other than a conversation that displaces a focused and individual succession of ideas. Toby and Walter's language produces a concert of exchanges—"a train of artillery," "a train of a fiddle stick"—which may not represent the meeting of minds, but does keep them together as they while away the tense duration of Tristram's birth. And from this perspective, we can recognize one last doubleness in the chapter: Walter's dissertation on duration, while being on topic, is off beat. It interrupts the moment's rhythm of dialogue; accordingly, Toby's interjections, while not attached to ideas, aim to nudge Walter back onto the pulse of exchange.

Throughout the novel Toby and Walter's bond rides on long conversations that feature a dissensus of opinion along with strong emotional attachment that arises from communication of shared beats rather than shared meaning. Though they almost always disagree, the sounds and gestures of their exchanges produce harmonious concerts and rhythmic synchronization. Take, for example, their argument about the strain of childbearing on women:

> —'Tis a heavy tax upon that half of our fellow-creatures, brother *Shandy*, said my uncle *Toby*—'Tis a piteous burden upon 'em, continued he, shaking his head.—Yes, yes, 'tis a painful thing—said my father, shaking his head too—but certainly since shaking of heads came into fashion, never did two heads shake together, in concert, from two such different springs.
>
> God bless } 'em all—said my uncle *Toby* and my father,
> Duce take } each to himself. (1.340)

Headshaking provides an isochronic beat for the conversation around which other sonorous elements harmonize. A hissing string threads the exchange from the recurring "'tis," "tax," "piteous," and "yes, yes" and onto "bless" and "duce," culminating in rhythmic synchronization of monosyllabic words with opposing meanings. Walter's "Duce take" is characteristically of harsher sound—produced by tongue and teeth—than Toby's "God bless," but the monosyllabic words used by both underscore the beat they have been sharing at least since their headshaking began. That the final sentences may either be quietly voiced by the brothers or silently rehearsed in their heads underlines the extent to which the rhythmic bond permeates subjectivity; Walter and Toby may have very different feelings about women, but as they partake in a conversation even their opposing private thoughts synchronize in the unvoiced form of their acoustic sound images. Moreover, that Sterne renders this synchronization in special typeface suggests that, like Steele, he wants readers and not just present speakers to experience this rhythmic synchronization and, through it, the shared sonorous duration of this conversation. It would be impossible for us to read Walter and Toby's lines in chorus if we follow customary practices of solitary novel reading; but through innovative typeface

Sterne does his best to indicate the shared experience at stake. And if we cannot make Walter and Toby's lines overlap in solitary reading, our own experience of the sound images doubles the rhythmic effect, iterating twice the duple metrical cycle—"God bless 'em all" "Duce take 'em all."

Sterne relentlessly emphasizes the importance of beats and often indicates that these not only rhythmically structure *Tristram Shandy* but also generate its very prose. He may or may not agree with Steele and with early musicologists that rhythmic sound strings tap into physiological vital pulses, but he nonetheless makes recurring beats an essential ingredient of his prose.[29] Indeed, pulsations seem like the very engine of writing for this novel's distinct mode of verbal sprawl. Tristram argues, for example, that not new sights, but the equable motions of riding, generate travelers' effusive writing. And if regular beats rather than novel knowledge produce discourse, then authors may just as effectively pace in their own backyards:

> "Now before I quit Calais," a travel-writer would say, "it would not be amiss to give some account of it."—Now I think it very much amiss—that a man cannot go quietly through a town, and let it alone, when it does not meddle with him, but that he must be turning about and drawing his pen at every kennel he crosses over, merely o'my conscience, for the sake of drawing it; because, if we may judge from what has been wrote of these things, by all who have *wrote and gallop'd*—or who have *gallop'd and wrote*, which is a different way still; or who for more expedition than the rest, have *wrote-galloping*, which is the way I do at present——from the great Addison who did it with his satchel of school-books hanging at his a—and galling his beast's crupper at every stroke—there is not a galloper of us all who might not have gone on ambling quietly in his own ground (in case he had any) and have wrote all he had to write, dry shod, as well as not. (2.579–80)

Refusing to report on Calais may suggest that in the dialectic of diurnal form, as Stuart Sherman calls it, Tristram prefers selectivity over exhaustive reporting, at least in the genre of travel writing if not in the genre of novel writing. Yet Tristram's digression into an argument about pulses as a writer's muse also indicates his commitment to a connection between beats and language.[30]

Even more, Sterne devises a breathtaking range of techniques for rendering beats. This range includes invocations of ticking clocks and pendulums, raps, ringing bells, pulsating hearts and arteries, galloping horses and pacing humans, thematic discussions of repetition, and iterative plot structures, as well as a sonorous prose of refrains, monosyllabic word strings, paratactic syntax, and regulated pauses made visible in print through commas and dashes. Dr. Slop's damning of Obadiah with Ernulphus's curse exemplifies many of these techniques of pulsation. The emotional intensity of the scene is drummed along by a pattern of refrains made of repeating syntactical structure and framed by sets of commas and dashes. Here is the first iteration:

> May the Father who created man, curse him.—May the Son who suffered for us, curse him.—May the Holy Ghost who was given to us in baptism, curse him (Obadiah.)—May the holy cross which Christ for our salvation triumphing over his enemies, ascended,—curse him. (1.205)

This rhythmic pattern recurs in four consecutive stanzas with two differences: in the last instance refrains of "damn him" replace "curse him," and in each instance a growing list of cursers expands the syntactical string ("—May all the angels and archangels, principalities and powers, and all the heavenly armies, curse him" [1.207], runs the second stanza). The growing list of cursers intensifies the oath both conceptually—enlarging Dr. Slop's supporting chorus—and musically—as the additional commas that punctuate the list add extra beats. After four such rounds, the curse proceeds to a different stanzaic form made of paratactic lists punctuated by commas:

> May he (Obadiah) be damn'd where-ever he be,—whether in the house or the stables, the garden or the field, or the highway, or in the path, or in the wood, or in the water, or in the church.—May he be cursed in living, in dying. (1.207)

This stanza also, with its high incidence of commas, continues to underline pauses as a key form of linguistic beat, a form further emphasized by the grand finale of the oath, made up of another very long list of curses.[31] In this curse, not poetic meter—a recurring pattern of equal numbers of syllables and parallel positions of stresses—but recursive beat rendered through pauses, parataxis, and refrains, constitutes rhythm and its emotional import.

Sterne continues in his next chapter to suggest that the force of Slop's affecting performance rests in recursive structure, especially a recursive structure of pauses. A narrational interlude available only for readers between two events of plot—Slop's passionate swearing and Susannah's excited report of all the mishaps at Tristram's birth—the chapter features Tristram's contention that no curse is an original. If the chant dramatizes a sonorous recursive structure, then Tristram's narration conceptually underlines iteration by claiming that it structures oaths more generally insofar as any curse is always a rendering of one that has already been performed.[32] But as soon as Tristram declares that "I'll undertake this moment to prove it [that no curse is an original] to any man in the world" (1.212), he digresses into a dissertation on elocution:

> ——And how did *Garrick* speak the soliloquy last night?—Oh, against all rule, my Lord,—most ungrammatically! betwixt the substantive and the adjective, which should agree together in *number*, *case*, and *gender*, he made a breach thus,—stopping, as if the point wanted settling;——and betwixt the nominative case, which your lordship knows should govern the verb, he suspended his voice in the epilogue a dozen times, three seconds and three fifths by a stop-watch, my Lord, each time.———Admirable grammarian!———But in suspending his voice——was the sense suspended likewise? Did no expression of attitude or countenance fill up the chasm?—Was the eye silent? Did you narrowly look?—I look'd only at the stop-watch, my Lord.——Excellent observer! (1.213)

That no curse is an original may seem to have nothing to do with Garrick's performance unless we recognize that Garrick's idiosyncratic elocution features, just like Slop's oath in the preceding chapter, a recursive rhythm. Garrick performs his speech with pauses that defy grammatical rules, but that are isochronic, Tristram tells us. These pauses may or may not enhance meaning—Tristram, like Tim Page listening to Glenn Gould, is too busy timing the performance to be able to tell its conceptual content. Garrick, however, seems to know that capturing his audience's attention and drawing its sympathy involves endowing his speech with a rhythmic pulse.[33] Similar to the passages from Smith's *The Theory of Moral Sentiments* we looked at earlier, Sterne suggests that even theatrical sympathy, when it is at its best, relies on shared rhythmic durations.

Ernulphus's curse and Tristram's short dissertation on elocution, then, reveal the utility of recurring beats for affecting performances—in secular theaters of soliloquies as much as religious rituals of curses. But with scampering discourse, recurring beats, and shared rhythmic durations, Sterne also makes a point about the potential efficacy of novel form—not just conversations and performances geared to drum up shared sympathetic moments in presence, but also narration in print designed to forge shared qualitative time across geographical and temporal distances. Within the sheer range of techniques Sterne uses to render pulses, several—like commas and dashes—are specific to writing. Janine Barchas explains that such typographical marks were commonly used in the literature of sensibility to "visually document[s] auditory and temporal aspects of 'real' speech which cannot be adequately captured by verbal transcription alone."[34] Sterne, as we've seen, makes ample use of dashes and commas as well as less common typeface for rendering rhythm and synchronizing beats, thus conveying sympathetic moments across the divides of diegetic communities and readers—a divide that is not only spatiotemporal but also ontological. In rendering beats through typeface as well as through sonorous prose, Sterne mechanically reproduces sympathetic duration in readers—a collective experience on the receiving end, even in the absence of actual bodies and subjectivities on the originating side. It is the sympathy generated not by another human being (as Garrick's special theatrical presence may misleadingly indicate), but by rhythmic form—musical, linguistic, and novelistic.

John Mullan argues that "Sterne concedes that sentiment can only be glimpsed across the distance between a translator and an 'original'; that while feeling is supposed to transcend words, it takes words (at once judicious and inaccurate) to translate sentiment"; and he concludes that "*Tristram Shandy* is writing away from the moment."[35] By this account, sentiment's particularity and qualitative fullness eludes the abstraction of language, while also requiring language for its communication. Similarly, Jonathan Lamb points out: "Confusion, sudden disappointment, and death seem to cause either a loss of words or a torrent of them. . . . In the latter cases, the words are usually remote from what they want to handle, especially when they are addressed vicariously to a sorrow in the form of consolation. Time and again in Shandean crises, words of comfort fail to touch the 'affection' of grief they are meant to eradicate."[36] But Lamb continues to propose that despite such moments of linguistic failure, Sterne's novels convey a manner in which words and experiences rely on similar associationist principles and, hence, are more continuous with one another than such moments of extreme emotion may suggest. Lamb explains that both Hume and Hartley "are agreed that all thought is a species of sensation" and that "there are very few situations in Sterne's two novels that don't develop a reciprocal play of sensations and ideas corresponding either to Hartley's scheme of vibrations or Hume's of gravitational pull."[37] Hartley posits a material link between words and ideas, maintaining that initial sensations are stored in the brain as continuous miniature vibrations that, in turn, evoke words. Hume, by contrast, posits a relation of analogy; not mechanical links, but intellectual attraction—belief—ties words to material experience. Yet by both logics, adopted and developed by Sterne, words mediate—whether by force of material or by force of belief—physiological experience as much as they mediate abstract meaning. Furthermore, according to Lamb, words at their best in Sterne's writing convey both.[38] To Lamb's flagging of Hume and Hartley's associationism as underwriting Sterne's project I have added the sensationist logic of early musicology and elocutionary theory. *Tristram Shandy*, I have been arguing, produces the sensible moment as sympathetic durations felt and conveyed through rhythmic language, in line with contemporary understandings of musical time and linguistic rhythm that underline beat as the essential dimension of rhythm and its sympathetic effects.

But there is a second dimension of Sterne's rendering of duration—his careful crafting of a doubleness that frames sympathetic moments with linear progression. On the one hand, *Tristram Shandy*'s narration renders and even amplifies the rhythms of diegetic conversations, thus conveying its durational fellow-feeling to readers; on the other hand, it underlines the difference—and specifically the chronological distinctions—between readers and diegetic communities. At stake is pointing out that while fictional characters may fully indulge in dilated sympathetic moments, real people—readers—can never fully extricate themselves from the relentless linear progress of chronological time. Individual endurance, this reminds us, is not only qualitative but also finite. The scene of Le Fever's dying serves as an especially clear example of such doubleness of narration—highlighting at once the linear distance that readers can never fully forego and the rhythms of sympathetic moments that characters and readers, nonetheless, share.

Faced with Trim's conviction that "the lieutenant's last day's march is over" (2.508), Toby insists on nursing the dying man so as to "set him upon his legs" (2.510). Here again Sterne presents a conversation in which passionate attachment and rhythmic synchronization accompany ideational disagreement:

> ——In a fortnight or three weeks, added my uncle *Toby*, smiling,—he might march.
>
> ——He will never march, an' please your honour, in this world, said the corporal:——He will march; said my uncle *Toby*, rising up from the side of the bed, with one shoe off:——An' please your honour, said the corporal, he will never march, but to his grave:——He shall march, cried my uncle *Toby*, marching the foot which had a shoe on, though without advancing an inch,—he shall march to his regiment.——He cannot stand it, said the corporal;——He shall be supported, said my uncle *Toby*,——He'll drop at last, said the corporal, and what will become of his boy?——He shall not drop, said my uncle *Toby*, firmly.——A-well-o'day,—do what we can for him, said *Trim*, maintaining his point,—the poor soul will die:——He shall not die, by G—, cried my uncle *Toby*. (2.510–11)

Toby's refrains of "he might/will/shall march" rhythmically punctuate this dialogue, and Trim's recurring "he will never march"—only slightly revising Toby's phrase—further underlines the conversation's iterative form. Toby's own attempts to march as he voices his faith in Le Fever's recovery add the pulse of attempted pacing, lending emphasis to the rhythmic quality and emotional import of the language. Tristram's narration interlaces its own beat with that of the characters' dialogue through recurring speech attributions—"said my uncle *Toby*," "said the corporal"—and through recurring long dashes, here used to visually mark transitions between speakers. Yet Tristram's comment that Toby marches "without advancing an inch" begins to carve an ironic perspective over this sentimental scene; it suggests that sonorous duration ought to be considered not only for its absorptive capacities but also for how it stands in relation to time's linear progress. Indeed, Toby's successful immersion in present hope fails by the standards of probable calculations: Trim is right—Le Fever will soon die.

Sonorous duration's immersive hope and objective time's ironizing perspective continue to jostle the next morning as Toby tries to pull Le Fever into his optimistic sense of the moment. Le Fever's landlady tells Trim, "I heard the death-watch all night long" (2.504), but Toby responds to these symbolic beetle-ticks with an especially vigorous scampering of discourse. He marches unto Le Fever's bed, immediately launching a series of solicitous questions and "without giving him [Le Fever] time to answer any one of the enquiries, went on and told him of the little plan which he had been concerting with the corporal the night before for him" (2.512). Toby communicates the emotional import of his words so well that "before my uncle *Toby* had half finished the kind offers he was making to the father, had the son insensibly pressed up close to his knees, and had taken hold of the breast of his coat, and was pulling it towards him" (2.512). But if this speech forges a strong emotional bond among the scene's characters, Tristram's narration continues to recall the perspective of time's advancement. Already with Toby's "concerting," Sterne flags the doubleness of the scene: on the one hand, "concerting" as in musically collaborating or performing, the kind of rhythmic attunement he associates with sensible duration, and on the other, "concerting" as in contriving or arranging, the kind of planning underwritten by an understanding of probable chronologies. Furthermore, as Tristram continues to narrate the scene, he conjures moments of sympathy for readers in rhythmic language and frames the sonorous effects of narration with ironic meaning. Here is how he concludes the chapter: "Nature instantly ebb'd again,——the film returned to its place,——the pulse fluttered——stopp'd——went on——throb'd——stopp'd again——moved——stopp'd——shall I go on? No" (2.513). The paratactic phrasing together with recurring dashes mimic the beat of Le Fever's last pulses, thus communicating to readers the rhythmic structure of the moment. And the final "No" in response to the question "shall I go on?" amplifies the lieutenant's terminal beat—Le Fever is no more—while semantically anticipating the first words of the following chapter, by which what has run out is not Le Fever's life but, rather, Tristram's patience. "I am so impatient to return to my own story," Tristram declares (2.513).

Invoking the imperative of marching on with plot, Tristram indicates that, in the final analysis, scampering discourse fails to aver death. He thus casts Toby's immersion in sonorous duration as a futile denial of inevitable realities.[39] Yet in persisting to represent such sonorous duration and in making a point of drawing in readers through pulsating prose, Sterne underscores narration's function of conveying qualitative duration. Narration in *Tristram Shandy* clearly differentiates readers from characters at the same time that it draws the reader into the rhythms of the diegesis. Carving for readers an ironic perspective, Sterne underscores their distance and difference from characters—we can be skeptical of Toby's estimations while he is naively immersed in the moment's presence. But at the same time Sterne's narration deploys rhythmic prose that replicates the beats of characters' conversations, thus constructing a peculiar reading experience that functions as the kind of synchronized beating time to another's joys and sorrows that Smith invokes—even as these others are nothing but fiction. We cannot forget that human time has an end, yet we continue to feel its fullness.

This doubleness of distance and immersion produced by Sterne's narration delivers yet another insight about human endurance: insofar as narration conveys qualitative duration to readers anywhere and at any time, it suggests that even as qualitative duration is finite on the level of the individual, it is infinite on the level of our species. As long as there are readers who can get Sterne's rhythmic prose—who can read in the way Steele recommends, keying into their pulse and voicing and dancing if necessary—qualitative duration will continue to be felt and human time will continue to endure. Addison, Diderot, and Richardson, as we have seen in previous chapters, have all entrusted leisure reading with the task of connecting us to a peculiarly human qualitative duration, but for Sterne, it seems, the stakes are even higher. To read—and to get—rhythmic sympathetic prose is to participate in that specific form of time and narrative that guarantees the persistence of human time.

## The Duration of Narration

I have been arguing that Sterne designs his sonorous prose to convey rhythmic sympathy and to make a philosophical point about the humanity of qualitative duration and its persistence through narrative. I now want to suggest that with such pulsating prose Sterne also makes an important contribution to narratology—conceptualizing that which from the perspective of plot seems like digression, as a technique of narration charged with conveying narrative duration. Sterne's writing thus refigures romance digression into a novelistic tool for conveying the feeling of time, narrative temporality understood not as chronologies of plot but, rather, as rhythms of discourse.

*Tristram Shandy* famously puts discourse and action in competition, staging, as Thomas Keymer observes, a Scheherazade model whereby Tristram's digressive narration delays the getting on with the plot and thus also postpones the threat of finality. From this perspective, digressions produce a dilated temporality that only reveals the "insufficiencies of representation," as Keymer puts it; for even as the novel's discourse retards the set schedule of events, disease propels both narrator and author towards life's closure.[40] This is the familiar romance mode, as Patricia Parker articulates it, where narrative is structured by tension between the forward propelling teleological force of quest and digressive dilations that, while clearly holding value, cannot be perceived as anything but error. Along with such a model of discursive digression, which defines dilated temporality by the events that frame it and casts digression as delay, Sterne offers an alternate model that highlights an internal independent logic for dilated moments. This logic rides on the rhythmic form of sympathetic duration as well on the social value of durational sympathy. For Sterne, narration's capacity to convey qualitative time—sympathetic, embodied, and rhythmic—points out the extent to which the competition between action and discourse that *Tristram Shandy* stages ought not to be understood solely in terms of delay or error—analyzing digression for what it does to the action—but also in terms of anisochronies—analyzing the rhythms produced by the combination of action and discourse. Understood in this way, Sterne's musical discourse not only refigures romance dilation as narrative rhythm but also uses romance's principle of completion—that human connectedness that improbably ends romance, as Michael McKeon has observed—as its very vehicle of punctuation. Rhythmic sympathetic moments in *Tristram Shandy* neither advance the plot nor end it, but they do bind characters and readers together off and on all along the narrative.

In stressing the importance of anisynchronies for Sterne's project, I am also suggesting that Sterne anticipates Gérard Genette, who pointedly remarks that "a narrative can do without anachronies, but not without *anisynchronies*, or, if one prefers (as one probably does), effects of *rhythm*" (88; emphasis in the original).[41] In his chapter on duration, Genette raises the difficult problem of assessing what the duration of a novel might feel like for readers, given the infinitely diverse conditions in which novels are read. His solution is to analyze ratios of action and narration and, thus, to formalize duration as rhythm, which, he explains, conceives of novelistic time in terms of musical logic. Such ratios of lines or pages per event, according to Genette, can be grouped into four movements, "whose relationships of succession and alternation governed structures like those of the sonata, the symphony, or the concerto for some two centuries" (94).[42] This move beyond chronologies constitutes Genette's greatest contribution to narratology, even if subsequent scholarship continues to emphasize questions of plot sequence, as Michael Levenson protests.[43] And yet, Genette's insight might be pushed even further, as shown in the previous section where I demonstrated how for Sterne the alternation of action and narration is only one among a wide range of techniques that produce effects of rhythm. More specifically, I indicated the many ways in which Sterne's prose mobilizes microlevel linguistic effects of rhythm, thus according narration itself with duration—not a time that can be assessed chronologically or chronometrically, but a temporality that can be assessed qualitatively.

Recognizing that narrational rhythms can be used for imparting qualitative duration is a point worth making because Genette reaches an impasse when he attends to the duration of narration. He argues that subsequent past-tense narration always masquerades as durationless: "One of the fictions of literary narrating—perhaps the most powerful one, because it passes unnoticed, so to speak—is that the narrating involves an instantaneous action, without a temporal dimension. Sometimes it is dated, but it is never measured" (222). For Genette, Sterne's work marks a peculiar exception: "Telling takes time (Scheherazade's life hangs by that one thread), and when a novelist puts on his stage an oral narrating in the second degree, he rarely fails to take that into account. . . . Nevertheless—and this is finally very odd—the fictive narrating of that narrative, as with almost all the novels in the world except *Tristram Shandy*, is considered to have no duration; or, more exactly, everything takes place as if the question of its duration had no relevance" (222). And he concludes: "Contrary to simultaneous or interpolated narrating, which exist through their duration and the relations between that duration and the story's, subsequent narrating exists through this paradox: it possesses at the same time a temporal situation (with respect to the past story) and an atemporal essence (since it has no duration proper)" (223). Genette restricts his assessment of duration here to measurement of expanse. Surprisingly, he overlooks the way in which subsequent narration may underscore its own duration not as quantity but as quality—keyed not to clock but to rhythm.

I want briefly to demonstrate, by way of conclusion, that subsequent literary narrating has often been acutely aware of its potential for communicating qualitative temporality and has mobilized Sterne's technique for this purpose. Sterne's technique and understanding of narrative duration, rather than being idiosyncratic, exemplifies a dimension of the sensibility chronotope that we find in a long tradition of novels. Take, for example, the two consecutive narrational accounts of the final proposal scene of Jane Austen's *Sense and Sensibility*. The first presents the proposal from the point of view of the impersonal narrator. It uses a vocabulary of hours and impassionate reporting:

> How soon he had walked himself into the proper resolution, however, how soon an opportunity of exercising it occurred, in what manner he expressed himself, and how he was received, need not be particularly told. This only need be said;—that when they all sat down to table at four o'clock, about three hours after his arrival, he had secured his lady, engaged her mother's consent, and was not only in rapturous profession of the lover, but in the reality of reason and truth, one of the happiest of men. (306)[44]

These clean sentences draw attention to events and their scheduling in a steady prose that, to the extent that it not only tells of Edward's three hours but also conveys a particular durational experience, is of rational measure—a thin, minimally qualitative temporality. The second presentation of the proposal scene, by contrast, achieves a remarkable feat of combining free indirect discourse with narrational telling, giving a summary account of Elinor inhabiting this time while also conveying the emotional urgency and momentary durational feel of her experience:

> But Elinor—How are *her* feelings to be described?—From the moment of learning that Lucy was married to another, that Edward was free, to the moment of his justifying the hopes which had so instantly followed, she was everything by turns but tranquil. But when the second moment had passed, when she found every doubt, every solicitude removed, compared her situation with what so lately it had been,—saw him honourably released from his former engagement, saw him instantly profiting by the release, to address herself and declare an affection as tender, as constant as she had ever supposed it to be,—she was oppressed, she was overcome by her own felicity;—and happily disposed as is the human mind to be easily familiarized with any change for the better, it required several hours to give sedateness to her spirits, or any degree of tranquility to her heart. (308)

Here the vocabulary of moments and instants along with the dashes, the commas, and the very long final sentence convey a temporal experience far richer than the matter-of-factness conveyed in the first passage. If the first passage reminds us that we all—characters and readers—inhabit the same homogenous empty time (even if the ontological statuses of characters and readers differ within it)—the second passage invites us briefly to inhabit Elinor's peculiar durational experience. This is not a naïve realism that imagines novels making it possible for readers and characters to inhabit durations so fully similar that they can count as the same—a point Austen clearly makes through the passage's unmistakable summary mode—but a virtuoso implementation of a novelistic technique that enables readers to get a qualitative sense of duration—the experiences of love and anticipation—through the very rhythms of narration.

It has been difficult for literary scholars to recognize that narration might impart various qualities of time, rather than just chronometric indexes, because the tools of narratology are not calibrated for such a task. Geared to pick up on the competition between action and narration—large swaths of prose in their representational capacity (whether of commentary or of events)—narratology cannot gauge the pulses of narration, the rhythms of words and sentences in and of themselves. Genette's insights about the importance of readerly experience and the usefulness of musical paradigms have not been pursued far enough. Not even by Genette himself who, in *Narrative Discourse Revisited,* qualifies his claims regarding the subsequentness of past-tense narration without reconsidering its duration and how this might be shaped by the effects of rhythm in the narrating instance.[45] Garrett Stewart addresses this limitation of traditional narratology by proposing a method designed to pick up on microlevel effects of rhythm, which he calls narratography. In contrast to traditional narratology, narratography attends to sentence-level poetic effects; it is concerned with the very selection and combination of words in the syntactical strings of narration and their various relations to plot. It "ask[s] less what *makes* a narrative than what *makes it tick* over syntactic time," as Stewart explains in his recent narratographic study of Victorian novels.[46] Stewart's analysis picks up on the ways in which novelists from Dickens through Hardy render duration through the very sonority of their prose, evidencing that Sterne's work with durational narration has been continued by a long tradition of novelists—running at least as far as Victorian high realism, according to Stewart's study (though the case could also easily be made for modernism).

When Stewart juxtaposes microlevel linguistic effects of rhythm with chronologies of plot, he consistently finds terror. In his account, Victorians connected moments of stylized prose with the pervasive violence of novels' plots. Thus, for example, Stewart argues that in Dickens's *Little Dorrit*, the sonorous prose of the last chapter points to the violent repressions of Arthur Clennam's biological mother and of Little Dorrit's doubling as both this disappeared woman and her banisher. Moreover, Stewart continues, this repression evidenced in the prose enables the neat marital closure of the novel's plot. I have suggested, however, that Sterne's stylized prose engages the stories it tells very differently. Tristram's narration does not aggressively expose Toby's misrecognition of the pulse that signals impending death; instead, it doubles irony alongside the rhythmic-sympathetic bonds that this pulse produces. Neither plot nor sonorous prose violates character, but both produce only a slightly more knowledgeable perspective over characters' immersion in the all-too-human.[47] But regardless of these differences in the ethical import of content, Stewart's study of Victorian fiction indicates that Sterne's approach to narration's duration stands not as an exception to the rule, but as the instance that sets the rule. In explicitly coupling his scampering discourse with effects of rhythm, Sterne establishes a key aspect of the sensibility chronotope whereby musical time underwrites the temporal dimensions of narration and conveys qualitative duration to readers.

*Chapter 4*

# Durational Aesthetics and the Logic of Character

## Radcliffe, Burke, and Smith

At the Live Oak Quaker meeting house in Houston, Texas, James Turrell installed a roof piece, which opens for the public every Friday at dusk. One walks in, sits down on one of the benches, and looks up through a sharp-edged, perfect square window in the roof, observing the process through which the light blue sky gradually transforms into black. It is a meditative experience—a break from the ordinary rush of daily life. But Turrell underlines this break not as a moment of stillness, not exactly as time-out, but rather as slow time or a stretched-out moment, in which the external, ongoing but nearly imperceptible process of dusk promotes an internal, ongoing but nearly imperceptible process of thinking and feeling. Framing dusk, Turrell asks us to attend to the processual nature of our attention and to the duration of our aesthetic experience.

Turrell might be understood to literalize an evening topos familiar in a long poetic tradition. Christopher Miller identifies this tradition as "Romantic time," evolving from Virgilian and Miltonic tropes into a "nocturnal *sensibility* associated with the sonnet revival of the 1780s and 90s (what is understood, in terms of literary periodization, as Sensibility, *tout court*)" (8) and peaking in early nineteenth-century English verse.[1] In its lyric form, and especially in its Romantic apotheosis as Miller understands it, the evening topos serves to mark a poetic moment as one in which the durationality of thought, feeling, and language are intensely felt; it also serves to recognize this moment as a break from the ways in which thought, feeling, and time are experienced ordinarily; and, finally, it serves to acknowledge "a moment in which poetic reflection becomes a tertiary act between secular rest and sacred worship" (9). But if this poetic tradition insists on delivering evening and the reflectivity it may promote as the domain of a specialized representation that activates specialized imaginative capacities—a lyric topos in conversation with its poetic precedents and promoting not simply consciousness but the self-consciousness of a tradition—then Turrell's roof piece suggests that merely framing dusk would connect anyone to the durationality of pleasure and reflection. This may be taken as a reductive, naïve, or disingenuous attitude toward representation—stripping it of its representationality and the discursive traditions that constitute it. Nonetheless, it underlines a recognition of the wide range of interventions that might be mobilized to achieve a consciousness of duration. Not just lyric poetry, but also installation art, and—as we will see in this chapter—meandering and spiraling lines, music, and prose fiction can all be configured to produce such markedly durational aesthetic experiences. The mere framing of dusk also enables us more easily to recognize this durational aesthetics as a contribution to the phenomenology of time. Released from dense discursive allusiveness, it flags time and its experience as its sole focus, directing us less to consider how it stands in relation to other representations and more to how it stands in relation to our temporal perception.

It is this more broadly defined durational aesthetics that the sensibility chronotope relies on. Mid- and late eighteenth-century writings often articulate aesthetic experience as both crucially temporalized and as a means for intensely feeling the very temporality of consciousness. These writings directly engage John Locke's theorization of duration as defined by the succession of ideas and of the instant as defined by immeasurable sensible experience, extending and more fully developing what I have identified in Chapter 1 as Joseph Addison's associationist aesthetics. Though aesthetic writing in the eighteenth century is certainly concerned with judgments of taste, at stake in the approaches I track in this chapter is less the atemporal justification of preferences—though obviously the writers I consider here also do that—and more the psycho-physiological processes that aesthetic enjoyment relies on. Such focus on perception and what it feels like enables these aesthetic theorizations and practices to contribute to representational traditions as well as to the conceptualization of time.

As we saw in Chapter 1, Addison identifies two styles of writing—nonmethodical and methodical composition; and though he clearly judges the methodical composition as better, his discussion focuses on how each activates different forms of reflection and supports a different sense of duration. The nonmethodical essay is composed as a meandering line and elicits pleasing confusion; the methodical essay is composed more like a spiraling line looping memory and anticipation as it also proceeds steadily and promotes reassuring command. In this chapter I examine how this conceptualization of lines with their temporalities and pleasures develops in durational aesthetics ranging from discussions of visual arts, to music, to fiction. Edmund Burke in his *A Philosophical Enquiry into the Origin of Our Ideas of the Sublime and Beautiful* (1757) picks up the pleasures of meandering and of the sensible moment. Adam Smith in his theorization of instrumental music in "Of the Nature of That Imitation Which Takes Place in What Are Called the Imitative Arts" (1795) develops the pleasures of the spiraling line and the more orderly feeling of time passing. And Ann Radcliffe in *The Mysteries of Udolpho* (1794) takes up both forms of lines and of durational aesthetics.

If Burke and Smith have reputations as interesting and sophisticated thinkers, Radcliffe does not. Radcliffe's reputation continues by and large to rest on her commercial success, and her craftsmanship is usually assessed for how it exemplifies eighteenth-century fashions and anxieties.[2] I argue, however, that Radcliffe's schematic landscape descriptions and her flat characterization, much like Turrell's framing of dusk, underline the temporal constitution of the medium she uses—the inherent successiveness of her linguistic art-form in its peculiar suitability to the durational consciousnesses that she aims both to represent and to affect. Radcliffe, like all the other writers I examine in this book, underlines duration as a peculiarly human experience of mental crafting, rather than as marking an elevated sensibility of a few privileged individuals. Typicality in Radcliffe is, thus, a program, rather than deficient art—a program designed to highlight the associationist temporal stakes of her representations.

But even as I argue that Radcliffe engages a broad discourse of durational aesthetics, my discussion also recognizes how she revises a more narrowly defined literary tradition of prose fiction. For another way to understand the typicality of her representations is to consider how her technique highlights what is sensible about her romances—which is to say what makes them refigurations, rather than imitations, of earlier instances of the genre. In his second preface to *The Castle of Otranto* (1764), Horace Walpole contends that the gothic form he inaugurates combines old romance plots in their improbability with new romance characters in their verisimilitude of human psychology. And yet a Manfred or a Matilda is hardly a Clarissa or even a Tom Jones; Walpole attends to his characters' consciousnesses only in summary. But if Walpole's characters seem closer to old romance types than to modern renderings of fictional persons, then Radcliffe suggests that this is because they insufficiently feature process, rather than insufficiently feature particularizing detail. In Radcliffe's fiction, as in chivalric romance plots, " 'suddenly' is normalized . . . it becomes something generally applicable, in fact, almost ordinary. . . . The unexpected, and only the unexpected, is what is expected," as Mikhail Bakhtin puts it.[3] But unlike this old romance logic, these adventures are grafted on the personalities of characters—not by way of content-laden thematic features, but rather by way of the associationist activity of crafting these random events into coherent biographical wholes, an activity which her heroes and heroines relentlessly perform. Thus Radcliffe, just as much as Samuel Richardson and Laurence Sterne, engages both the philosophical and the literary discourses that constitute the sensibility chronotope.

## Burke's Sensible Moments

In his *Enquiry,* Burke implicitly engages what Locke had described as events that are either too quick or too slow for us to register as successive. In this Burke implicitly takes up Locke's second definition of the instant as a part of time so as more carefully to account for its qualitative durationality. A fired bullet, to recall one of Locke's examples, hits the walls of a room and parts of the body, thus producing successive noise and pain, but human minds cannot register such rapid blows as discrete entities; at the other limit of human ideational pace, Locke explains, there's a clock, whose hands move too slowly for us to perceive their changing positions. These examples prompt Locke to revise his definition of the instant so that it includes the temporality of imperceptible succession. "Such a part of Duration as this, wherein we perceive no Succession, is that which we may call an *Instant*; and is *that which takes up the time of only one Idea* in our Minds, without the Succession of another, wherein therefore we perceive no Succession at all."[4] Developing this second of Locke's definitions of the instant, Burke underlines the processual nature of moments of intense experience, making the case for aesthetic pleasure as a duration constituted by our registering a sensationist—idea-less—succession. This is not quite the succession of physiological pulses taken up by the early musicology I examined in Chapter 3. Burke focuses on sensation not in order to discover the physiological underpinnings that can explain how our primary sense of duration may be constituted as equable flow. Rather, his purpose is to highlight the physiological processes that make any moment of experience temporal.

Burke begins his discussion of the sublime effects of infinity by underlining differences between objects and our perceptions of them: "There are scarce any things which can become the objects of our senses that are really, and in their own nature infinite. But the eye not being able to perceive the bounds of many things, they seem to be infinite, and they produce the same effects as if they were really so" (73).[5] Infinity is thus defined as an optical illusion—not a feature of objects, but an effect of our perceptual capacities. Alan Richardson points out the similarity between sublime experience and optical illusions, arguing that Burke examines how our brain shapes perception, rather than passively receives stimulation. Richardson thus recasts what earlier scholars have understood as Burke's unself-conscious sliding from empiricism to idealism into a self-conscious probing of the ways in which sensible external data interacts with subsensible impositions of the brain.[6] Joining Richardson's revisionary approach, it is also worth noticing the extent to which Burke's interest in primary levels of perception leads him to highlight duration and to privilege aural and tactile models of sensation over visual ones. Thus, if the first paragraph of the section on infinity considers objects that elude our vision, the next paragraph shifts from misperception to sensation and from vision to sound: "After a long succession of noises, as the fall of waters, or the beating of forge hammers, the hammers beat and the water roars in the imagination long after the first sounds have ceased to affect it; and they die away at last by gradations which are scarcely perceptible" (73). What the first paragraph illustrates as a spatial property of objects, the second paragraph highlights as a temporal feature of sensation which is best exemplified by aural stimulation. And the temporality of such sound gets further underlined by having no clear sense of an ending.

As Burke continues in the next chapter to detail this artificial infinity, he spells out the importance of idea-less succession in the experience of sublimity:

> Succession and *uniformity* of parts, are what constitute the artificial infinite. I. *Succession*; which is requisite that the parts may be continued so long, and in such a direction, as by their frequent impulses on the sense to impress the imagination with an idea of their progress beyond their actual limits. 2. *Uniformity*; because if the figures of the parts should be changed, the imagination at every change finds a check; you are presented at every alteration with the termination of one idea, and the beginning of another; by which means it becomes impossible to continue that uninterrupted progression, which alone can stamp on bounded objects the character of infinity. (74)

The bounded object, Burke explains, creates the illusion of unboundedness by accentuating the durationality of its perception—"continued so long . . . to impress the imagination with an idea of their progress"—but it does so precisely by *not* relying on a succession of ideas, for by such "termination of one idea, and the beginning of another . . . it becomes impossible to continue that uninterrupted progression." The artificial infinite gives us an experience of sublimity through a succession of sensations that yield only a single idea—that of ongoingness itself. Though it is not an equable beat—as it crucially is for early musicologists' conceptions of time—at stake is a recursivity that underscores the duration of experience, suggesting that sublimity enables us to register the successive constitution of the sensible moment.

Burke emphasizes temporality not only in his discussion of sublimity but also in his analysis of the beautiful, though he presents idea-less succession in the beautiful somewhat differently. Consider Burke's (in)famous passage about the beauty of a woman's bust: "Observe that part of a beautiful woman where she is perhaps the most beautiful, about the neck and breasts; the smoothness; the softness; the easy and insensible swell; the variety of the surface, which is never for the smallest space the same; the deceitful maze, through which the unsteady eye slides giddily, without knowing where to fix, or whither it is carried. Is not this a demonstration of that change of surface continual and yet hardly perceptible at any point which forms one of the great constituents of beauty?" (115). This example may tastelessly objectify women, but the emphasis on only "that part of a beautiful woman," which itself gets further parceled into metonymic seriality, helps clarify how aesthetic experience recasts an instant into a durational sensible moment. Tom Huhn notices that tactility here displaces vision—how the "insensible swell" may be "hardly perceptible" to the eye but should be intensely experienced in touch, thus suggesting that "The eyes are not, then, at least for Burke, the primary portals through which love and beauty enter us."[7] With such privileging of tactility, this example replaces an illusory atemporality with process; whereas vision gives the illusion of a whole object comprehended at once, touch senses its objects through successive reports of partial segments. Significant here is not only the seriality of tactile sensation in contrast to the synchronicity of vision, but also the impossibility of localizing points of change or identifying discrete segments—"the unsteady eye slides giddily, without knowing where to fix, or whither it is carried." If seriality underlines process, then the impossibility of localizing points of change emphasizes that it does so without registering succession of discrete ideas. The tactile intake of the bust's line models a special kind of durational dynamic—restless motion that intensely feels but cannot quite comprehend wholes and cannot deliver abstract concepts or whole images. Thus its duration is registered less as time in the way that Locke and, following him, Hume have defined it—the succession of ideas—than as a sensible moment.

Burke emphasizes "gradual variation" as an ideal of natural beauty more generally, arguing that the lines of birds and women "vary their direction every moment, and they change under the eye by a deviation continually carrying on, but for whose beginning or end you will find it difficult to ascertain a point" (114–15). He further stresses the importance of sensing gradation without locating discrete points of change by faulting William Hogarth for including angular variation in his examples of beautiful intricacy: "These figures, it is true, vary greatly; yet they vary in a sudden and broken manner; and I do not find any natural object which is angular, and at the same time beautiful" (115), Burke emphatically maintains. Hogarth introduces his waving and serpentine lines as "bewitching to the sight," giving us the sensation of dancing along in a country ball and occasioning for the eye a perceptual pursuit; an intricate line "*leads the eye a wanton kind of chase*, and from the pleasure that gives the mind, intitles it to the name of beautiful."[8] Hogarth thus seems as committed as Burke is to the experience of restlessness that gradual variation occasions, but his articulation fails to suggest a notion of idea-less succession. In flagging his polemic, Burke doesn't so much depart from Hogarth's aesthetics as emphasize the extent to which his own concerns lie with formulating the kind of idea-less duration that Locke's discussion of time could not quite account for—the duration of the sensible moment. If succession without change constitutes the sublime, then change that can be sensed even without points to index its successive stages constitutes beauty. Both are ways of experiencing succession without ideas. While the sublime may be occasioned by big and strong objects and the beautiful by small and weak ones, and while the sublime may promote solitude and self-preservation and the beautiful promote sociality and smugness, they are both experiences that make instants palpable: occasions for us to sense those durations whose being part of time we cannot otherwise register, insofar as we normally register time—according to both Locke and Hume—through the succession of ideas.

Recognizing the durational underpinnings of Burke's examples, their aural and tactile logics, and the claims they make for an idea-less succession can help us better understand the *Enquiry*'s turn to an antipictorialist poetics in its last section. This turn tends to puzzle twentieth-century readers, who find in it evidence for Burke's incoherent thinking about the relation between sensation, on the one hand, and symbolic orders, on the other. W. J. T Mitchell argues that the *Enquiry* offers two very different theories of the sublime: a moderate verbal sublime that will eventually align with Burke's later defense of British culture, and a raw sensationist sublime that rides on the visual feature of darkness and will eventually be repudiated as French.[9] But Mitchell's interpretation does not sufficiently acknowledge the extent to which the early sections of the *Enquiry* highlight situations in which ocular cues do not yield discrete whole images and, as such, these sections focus less on vision than on its failure. From the perspective of the readings of Parts I through IV that I have presented here, the turn to an antipictorialist poetics in Part V is less a negation of the treatise's sensationism than an extension of the case Burke makes for generalizable experience that is neither abstract nor predominantly visual, insofar as it attends to the durational nature of perception that both abstractions and vision elide. Much like the musicolinguistic conceptions of rhythm that I examined in Chapter 3, Burke's antipictorialist poetics, like his beauty and sublimity, aims to articulate a form—a pattern of communicable experience—that cannot be abstracted from the bodies that manifest and perceive it and thus counts duration as its essential quality. To a large extent, articulating a logic of form that cannot be abstracted from bodies was the hard problem tackled by eighteenth-century empiricists to the extent that they all agreed that knowledge arises from sense impressions. But Burke contests Locke's approach to this problem in an especially direct way, formulating his antipictorialist poetics against Locke's preference for prose language, which Locke values for conveying images and clear ideas.[10]

That all this is at stake in Burke's turn to language in the last part of the *Enquiry* might best be evidenced by the fact that he chooses to weigh in on the accomplishments of blind men, a subject often debated by eighteenth-century empiricists. Blindness becomes an especially useful test case for Locke and other empiricists because if our concepts arise only from sensory experience, then the data with which the blind man works to generate concepts differs considerably from that of the seeing man.[11] Burke considers the blind poet Thomas Blacklock and Cambridge's blind Lucasian Professor of Mathematics, Nicholas Saunderson. With Blacklock's example, Burke questions language's serviceability to ideas and promotes, instead, its effectiveness in communicating feelings. Blacklock is "a poet doubtless as much affected by his own descriptions as any that reads them can be; and yet he is affected with this strong enthusiasm by things of which he neither has, nor can possibly have any idea further than that of a bare sound; and why may not those who read his works be affected in the same manner that he was, with as little of any real ideas of the things described?" (168-69). Much like Sterne and the elocutionists, as we've seen, Burke argues that the blind poet and his seeing readers all enjoy language for its sound, a pleasure that has little to do with ideas and much to do with feelings. Indeed, Burke concludes the *Enquiry* writing, "All verbal description . . . conveys so poor and insufficient an idea of the thing described, that it could scarcely have the smallest effect, if the speaker did not call in to his aid those modes of speech that mark a strong and lively feeling in himself. Then, by the contagion of our passions, we catch a fire already kindled in another" (175). As in the examples of sublimity and beauty from the earlier sections, here, too, vision's failure points up aural and tactile sensations—the very sound of language and the fire of feeling—and the communication of something significant, but not of ideas.

The debate between the language of knowledge and the language of sentiment, which I considered in the previous chapter, helps make sense of Burke's comments on Blacklock. It does not, however, help to clarify his comments on Saunderson. For unlike the poet Blacklock, the natural philosopher

Saunderson is undoubtedly in the business of communicating intellectual concepts rather than sympathy. To understand what seems to Burke especially important about Saunderson's blindness, we must find a different explanatory framework. Saunderson was famous for his tactile arithmetic as well as for his lectures on color and light, accomplishments that Denis Diderot discusses extensively in his "Letter on the Blind" (1749). Saunderson, Diderot claims, may understand hypotheses about vision in a more immediate way than seeing men ever could, for he takes "a ray of light for a fine and elastic thread, or for a succession of minute bodies striking our eyes with incredible velocity, and he makes his calculations accordingly" (102). "Saunderson, then, saw by means of his skin" (107), Diderot concludes.[12] Burke does not flag the usefulness of Saunderson's predominantly tactile perceptual understanding, but he does indicate that Saunderson's blindness makes him much more acutely aware of language's special suitedness to convey the temporality of experience. That Saunderson could teach color to seeing men suggests to Burke that the effectiveness of language does not rely on invoking visual images stored in the mind, but on language's capacity to convey a liveliness—the way in which language highlights the processes constitutive of vital phenomena.

Burke's comment on Saunderson indicates that language can generalize and communicate process and its knowledge in a way that pictorial images do less effectively by soliciting modes of perception that are more sonorous or haptic than visual.

> He [Mr. Saunderson] did nothing but what we do every day in common discourse. When I wrote this last sentence, and used the words *every day* and *common discourse*, I had no images in my mind of any succession of time; nor of men in conference with each other; nor do I imagine that the reader will have any such ideas on reading it. Neither when I spoke of red, blue, and green, as well as of refrangibility; had I these several colours, or the rays of light passing into a different medium, and there diverted from their course, painted before me in the way of images. I know very well that the mind possesses a faculty of raising such images at pleasure; but then an act of the will is necessary to this; and in ordinary conversation or reading it is very rarely that any image at all is excited in the mind. If I say, "I shall go to Italy next summer," I am well understood. Yet I believe no body has by this painted in his imagination

> the exact figure of the speaker passing by land or by water, or both; sometimes on horseback, sometimes in a carriage; with all the particulars of the journey. Still less has he any idea of Italy, the country to which I proposed to go; or of the greenness of the fields, the ripening of the fruits, and the warmth of the air, with the change to this from a different season, which are the ideas for which the word *summer* is substituted; but least of all has he any image from the word *next*; for this word stands for the idea of many summers, with the exclusion of all but one: and surely the man who says *next summer*, has no images of such a succession, and such an exclusion. (169–70)

At some points within this passage, the disparity between language's generality and particular experience underwrites the difficulties of visualization—the "I" who is no image of anyone because it can by no means evoke "the exact figure of the speaker passing by land or by water, or both; sometimes on horseback, sometimes in a carriage; with all the particulars of the journey." And yet Burke uses "every day" as his first antipictorialist example not simply because it is an abstraction, but because it is, more specifically, a temporal abstraction—it refers to the "succession of time," for which we have no image. Similarly, "next" turns out to be strong evidence for the antipictorialist argument—"but least of all has he any image from the word *next*"—not only because it is yet to have a concrete existence, but also because it depends on the aggregation of a particular time of year—"the idea of many summers, with the exclusion of all but one: and surely the man who says *next summer*, has no images of such a succession, and such an exclusion." Burke thus explicitly grounds his antipictorialist linguistics in visual images' difficulty to convey time passing, a difficulty that language, just like aural and tactile perceptions, tackles more effectively.

Close attention to Burke's representation of summer in this passage indicates that here also—as in his examples of beauty and sublimity—he finds most compelling intense experiences that rely on incomprehensible sensorial succession, but in this passage such sensations are mediated by literary technique. Italy's next summer seems unvisualizable by virtue of the processual nature of the nouns and adjectives used to represent it. "Ripening" and "change" modify summer's fruit and summer's air respectively, attaching process to the static nouns. And if summer's greenness might be taken as an atemporal condition, then we should recall how at the beginning of this passage Burke makes the case for the blind Saunderson's ability to teach color by emphasizing the dynamic constitution of this phenomenon which vision cannot capture: "Neither when I spoke of red, blue, and green, as well as of refrangibility; had I these several colours, or the rays of light passing into a different medium, and there diverted from their course, painted before me in the way of images." Insofar as the successive states of light passing through a medium are essential to color and yet cannot be seen, a blind man can as easily describe it as a seeing man. In fact, Burke implicitly suggests—just as Diderot explicitly argues—that the blind man is better positioned to describe color precisely because he is not led astray by the atemporal illusion of the phenomenal image. It is such an understanding of the passage's trajectory that can explain how Burke may argue that when we speak of red, blue, and green we do not see them. Insofar as their visual existence relies on the illusion of traveling light being comprehended *not* through its successive refracted states but all at once, such images require a deliberate "act of the will." And if here Burke doesn't argue that the sensation of color per se has duration, his description is designed to give us the experience of the imperceptible duration of the traveling light of which it is constituted.

If a static image is inadequate to summer's changing from cool to warm air, or the process through which fruit ripens, then it also fails green's traveling light. Ripening and color are processes of different orders—the mediation of light is extraordinarily quick, noticeable only in experimental conditions or their discursive representation, whereas the transitioning of the seasons is often so slow that it is noticeable primarily in poetic exercises. But the two together parallel Locke's discussion of the flying bullet and hands of the clock—motions too fast or too slow to prompt succession of ideas, but which, nonetheless, evoke enduring experiences. The language of greenness and ripening—just like the sensations of beauty and sublimity of the earlier parts of the *Enquiry*—evokes durational experience that vision and discrete concepts or ideas fail to register. If in the earlier sections of the *Enquiry* Burke indicates that aesthetic pleasure—sublimity and beauty—is especially well suited to connect us with the durationality of the sensible moment, then in the last section of the *Enquiry*, and especially through the example of Saunderson, Burke argues that language (of all other media) is especially well suited to deliver this temporal aspect of aesthetic pleasure. Burke's discussion of Blacklock suggests that poetic language well conveys the temporality of sensation and feeling, but his discussion of Saunderson indicates that language as such, and not just poetry, is geared to convey such processes. Language can do this, in Burke's account, not because of its rhythms or sympathetic capacities, as the elocutionists and Sterne argue, but because it inherently narrativizes—always delivering succession, even when it represents a succession that we cannot otherwise perceive.

## Radcliffe's Nocturnal Sensibility

Burke works out, then, the durationality of the sensible moment, which—he believes—language more than other media is especially well suited to convey. In privileging linguistic representation and its conceptualizing of a durational moment, and in retiring vision and underlining haptic and aural sensation, Burke may be taken to inspire some of the nocturnal sensibility of late eighteenth-century lyric. Radcliffe contributes to this lyric tradition in *The Mysteries of Udolpho*, staging evening scenes in which the heroine's attention to a gradually darkening landscape compels her to compose verse, which Radcliffe includes as lyric breaks within the narrative prose. But Radcliffe's version of the evening topos underscores perceptual processes that recall Burke, more than poetic allusions that recall Virgil or Milton, and in her fiction the aesthetic moment traverses the visual break on the page between verse and prose, flagging the integration of the meditative moment into ordinary language and ordinary time, rather than identifying it as a specialized figuration of such language and time.

In the framing sensibility plot of *The Mysteries of Udolpho,* we often find Emily, the heroine, sitting alone on a mountain at dusk, viewing the surrounding landscape as its details become obscured by darkness, and being inspired to compose poetry. Toward the end of the novel in one such scene, twilight inspires Emily to write a "Song of the Evening Hour." The narration tells us that Emily, perched at a favorite spot, "linger[s] . . . to a late hour . . . watching . . . the gradual effect of evening" (509); then she composes twelve stanzas of rhyming quatrains in iambic pentameter that both personify and observe the process of twilight.[13] The poem draws attention to duration in various ways. The speaker tells us that she follows "that sweet, lingering strain of gayer Hours! / Whose close my voice prolongs in dying notes" (lines 9–10). She repeatedly emphasizes ongoing activity, casting conditions we customarily take as static into verbs that highlight process: herbs and flowers "all their fragrance on the air diffuse" (20); silver moonbeams "silver all the eastern cloud" (46). And the poem also underlines the progressive mode: "fading Day" (1), "lingering strain" (9), "dying notes" (10), "slumbering ocean" (16), "cooling dews" (18), "stretching plains" (23), "fresh'ning wind" (25), "Low breathing" (26), "Stealing its sweetness o'er some plaining rill" (30), "soothing" (31), "swelling" (32), "leading" (35). But both narration and poem document a process in which vision of distinct entities comes to be replaced by successive sensations. Emily observes the landscape "till the gray waters of the Mediterranean and the massy woods were almost the only features of the scene, that remained visible" (509), at which point she turns to poetry. As Radcliffe presents Emily's poem, readers transition from prose that communicates the fading away of an image to verse that intensifies the processual nature of the experience through rhythmic language and through the figuration of vision being displaced by other sensations. The speaker notes her "straining eye" (line 6) and as the "faint ray" (8) and "purple gleam" (15) fade into darkness, hearing, smell, taste, and touch come to dominate: the fragrance of herbs and flowers (19–20), the wind who "whispers soft" (27) and whose blowing of the reeds the speaker synesthetically "watch[es] to hear" (29). And right after the poem, narration returns to Emily, again emphasizing process and a transition from vision to other sensations, noting that she watches "the moon" "rising" and "its gradual progress" (510), and "as she sat meditating, sounds stole by her on the air" which she recognizes as the music of the ghost of Languedoc. These sounds evoke in Emily an "emotion of awe, which she felt, was not unmixed with terror" (510–11)—flagging the sublimity of her experience.

By the scene's end, Emily is late for dinner, exposing herself to being "gently reproached" by her surrogate father, the Count de Villefort. The indulgence in a view, which then turns into composing poetry, causes the dutiful Emily to be late for her engagements, as well as causes the committed fiction reader to pause her pursuit of plot. Because the poetry comes as the climax of the scene, Radcliffe may be taken to privilege verse in its capacity to figure such meditative durational experience and thus to support the notion of the evening topos as "Romantic time" in the way that Miller tracks it—"the emblematic time of lyric utterance. . . . At the end of the day, when other people have finished their work, the poet self-consciously announces the beginning of a different kind of labor—the invisible and silent activities of listening, watching, and thinking" (14).[14] And yet Ann Wierda Rowland points out that such privileging of verse is "a myth of poetry that has significant origins in the novel, staged by the insertion of lyric lines into the novel's narrative frame."[15] Moreover, if Radcliffe may be understood to contribute to such mythicization of poetry, it is only through the retroactive lens of the Romantic lyric; for in *Udolpho*, Emily's meditative break includes not only the character's verse creation but also the novel's descriptive narration. Radcliffe thus emphasizes that verse neither performs a specialized job nor enables a specialized experience that narrative cannot take up equally well.[16]

From the perspective of a Romantic individualism that prizes not just consciousness but self-consciousness, Radcliffe's landscape scenes might be criticized for neither sufficiently particularizing the scenery nor sufficiently individualizing the viewer. Marshall Brown argues, for example, that they count as superb examples of cliché; "they tell us little either about the vistas or about the psychology of the characters," he writes, even as they dramatize what Brown calls "the labors of description," by which he means an effort of projecting depth onto an essentially flat medium—language, gothic form.[17] For a Romantic, such laboriousness that results in stock figures seems like a lapse, but Radcliffe unapologetically announces that such typification is precisely what interests her. In the first chapter of *Udolpho* she declares that her landscape scenes are recurring motifs: "It was one of Emily's earliest pleasures to ramble among the scenes of nature," the narrator tells us, "in scenes like these she would often linger alone . . . till the last gleam of day faded from the west." The narrator then continues to detail their stock features: "Then, the gloom of the woods; the trembling of their leaves, at intervals, in the breeze; the bat, flitting on the twilight; the cottage-lights, now seen, and now lost—were circumstances that awakened her mind into effort, and led to enthusiasm and poetry" (9–10). Radcliffe emphasizes the sensationist durational makeup of these recurring stock moments: Emily's lingering at dusk yields a noticing of recursive succession—leaves "trembling . . . at intervals" and flickering lights "now seen, and now lost." And such lingering happens often—"one of Emily's earliest pleasures," and recurring throughout the pages of *Udolpho*. These scenes, in their very typicality as announced here, aim to capture not singularity, but a generic quality—like a window opening up to any portion of sky at any evening. Turrell's roof piece excludes all objects to make the generic aspect of such aesthetics so much easier to grasp. Radcliffe, by contrast, charts the full Burkean process of transition from discrete images to sensorial and emotional succession. These scenes aim neither to represent a particular landscape nor to represent an individual subjectivity, but to articulate—like Burke's *Enquiry* and like Turrell's roof piece—the fundamental durationality of perception, sensation, and reflection as such.

Later in this chapter I examine other ways in which *The Mysteries of Udolpho* features these durational processes of consciousness, their generic nature, and their alliance with novelistic—or what Radcliffe would have called sensible—characterization. But before doing so I want to indicate how Radcliffe uses Burke's ideal of gradual variation not only in the landscape scenes and for representing Emily's meditative pleasures, but also for more generally defining the visual hallmark of the gothic—its signature style of intricate lines drawn against the backgrounds of dusk and darkness.

Emily's first view of the castle of Udolpho serves as a good example. This scene affords the heroine no aesthetic pleasure—the Udolpho section of the novel includes no occasions for the heroine to indulge in natural beauty and compose poetry. Rather, this landscape description serves to introduce the reader to the atmosphere and symbolic code of the chapters to come.

> Emily gazed with melancholy awe upon the castle, which she understood to be Montoni's; for, though it was now lighted up by the setting sun, the gothic greatness of its features, and its mouldering walls of dark grey stone, rendered it a gloomy and sublime object. As she gazed, the light died away on its walls, leaving a melancholy purple tint, which spread deeper and deeper, as the thin vapour crept up the mountain, while the battlements above were still tipped with splendour. From those too, the rays soon faded, and the whole edifice was invested with the solemn duskiness of evening. Silent, lonely and sublime, it seemed to stand the sovereign of the scene, and to frown defiance on all, who dared to invade its solitary reign. As the twilight deepened, its features became more awful in obscurity, and Emily continued to gaze, till its clustering towers were alone seen, rising over the tops of the woods, beneath whose thick shade the carriages soon after began to ascend. (216)

At the cusp of entering the seat of mystery, Emily views it at twilight. Again failure of vision is rendered as a process with an especially palpable temporality, through changing light and color: light dying away, darkness gradually deepening, and obscurity becoming increasingly dominant. But this passage culminates (or nearly culminates, as we will soon see) with the awful contour of Udolpho castle, instantiating what Eve Sedgwick identifies as the gothic's privileging of line over color. Sedgwick interprets this linear aesthetics as the genre's commitment to structure at the expense of play and to the death instinct at the expense of both the pleasure and reality principles.[18] As such, this image serves adequately to introduce the episodes of the Udolpho portion of the novel, where anxieties of murder and conspiracy underwrite all incidents and determine all meaning. The duration that Emily and we as readers are made to feel so palpably at the threshold of the castle emphasizes the threat of impending death and casts the dynamic moment of twilight as a metaphor for life's progress toward its nullification, an inevitability that at Udolpho seems only to accelerate.

And yet Radcliffe carefully concludes this introductory passage with resumed linear motion—the carriage's ascent up the mountain, a feature which cannot be extricated from the vignette by its being articulated as a dependent clause in a sentence whose subject is the castle's increasingly "awful" "features" and "clustering towers." The carriage's linear motion up the hill clues us in to the possibility of action and thus of life—the possibility that line will not freeze into structure. Indeed, despite all expectations, few deaths actually occur inside Udolpho and rather than a suspension of duration we get chronometric disorientation, temporally marked by intense scenes of flight and emotional recollection. Within the imposing structure of the castle our heroine experiences not stasis, but a flurry of activity (albeit desperate) that ultimately promotes her survival, rather than dooms her to death.

In François Truffaut's gothic last scene of *Fahrenheit 451*, the refugees of the soulless, book-burning, glossily colored state survive among leafless, crookedly branched winter woods, walking serpentine paths as they recite the books they've memorized. The scene is ghostly, no doubt, but life and pleasure decidedly reside at these gothic-styled outskirts, while death, horror, and tyranny—inevitability without play—belong with the colorful geometric shapes of the metropolis. What I think Truffaut recognizes, and what Sedgwick overlooks, is how gothic style *opposes* intricate lines to structure, with the former representing the temporal dynamics of process, motion, and liveliness—like Burke's and Hogarth's lines of beauty—and the latter embodying the threat of timelessness. Within the ominous edifice of Udolpho, the heroine's task will be to keep the intricate line moving. Hence the flights within meandering corridors so characteristic of the genre. But for Radcliffe, even more important than such spatial visualizations are aural and discursive renderings, with which she characterizes the consciousness of her heroine. For this task she combines Burke's sensationist aesthetics of the sensible moment with another model of durational aesthetics, one committed to the succession of ideas, rather than to idea-less sensations, and that can be seen in eighteenth-century conceptions of musical composition.

## Composing Human Time in Smith and Radcliffe

In the same discussion in which Sedgwick perceptively yet incompletely considers the gothic's linear aesthetic, she also argues that the genre's adherence to convention highlights a degree zero of fictional character construction—identity in its structural condition of possibility, prior to sociocultural individuation. I want to pursue Sedgwick's superb point beyond its poststructuralist agenda. In the same way that gradual variation and intricate lines literally animate Radcliffe's landscape descriptions into durational consciousness, a foregrounded effort at fundamental associationist composition animates her characterization. But in order to gauge this, it is helpful to approach Radcliffe's characterization by way of a musical model that highlights the affinities between the form of music and human consciousness.

We saw in Chapter 1 that Hume analogizes melody to the succession of ideas and impressions that constitute our primary sense of duration, and Addison explains the pleasures of the methodical essay according to how it activates memory and anticipation. Mark Evan Bonds identifies in eighteenth-century musicology a similar concern with the affinities of composition and consciousness. In contrast to nineteenth-century musicologists who relied on spatial metaphors, for eighteenth-century writers, Bonds explains, "The instrumental work was seen as a wordless oration, and its form was viewed not so much as a harmonic or thematic plan but as an ordered succession of thoughts."[19] Adam Smith's discussion of instrumental music in his posthumously published "Of the Nature of that Imitation Which Takes Place in What Are Called the Imitative Arts" (1795) offers an especially good example of such relation between an associationist understanding of consciousness and the pleasure we derive from music, highlighting the temporality they both share.[20]

Smith's discussion, like Burke's *Enquiry*, often sounds as though it takes

Locke's account of duration as an implicit interlocutor, making art an occasion for experiencing the qualitative time that Locke's explicit arguments cannot quite articulate. But if Burke points out the insufficiency of a focus on ideas, privileges idea-less succession, and thus produces aesthetic experience as a sensible moment, Smith remains committed to coherent thought and focuses on varying patterns of the succession of ideas. Similar to Hume, Smith suggests that missing from Locke's definition of our primary experience of duration, and what instrumental music helps us recognize, is patterned differences among successions that yield varying qualities of durational experiences. And similar to Addison, such patterned succession gains its temporal underpinnings from prompting in its hearers the activities of recall and anticipation.

At the outset of his discussion, Smith contends that instrumental music ranks low in its capacity to imitate objects and motion. Yet this initial negative assessment enables him to recognize a strong affinity between minds and melody. Many good philosophers, Smith argues, mistake music's pleasure for mimetic achievement; Jean-Jacques Rousseau, for example, takes melody to imitate the human voice expressing passion. Yet, according to Smith, music evokes emotion not by representing a feeling person or telling a sentimental story, but by "becom[ing] itself a gay, a sedate, or a melancholy object," by which he means that "whatever we feel from instrumental Music is an original, and not a sympathetic feeling: it is our own gaiety, sedateness, or melancholy; not the reflected disposition of another person" (164).[21] Though we think we share emotion with music, feeling is only a byproduct of a more essential formal element of this art, which is also a formal element of human minds—their reliance on compositional principles that order succession.[22]

Music, by Smith's initial definition, is "a succession of a certain sort of sounds . . . regulated according to time or measure, and thereby formed into a sort of whole or system" (149), much like states of mind, as Smith explains, which are constituted by successive ideas regulated by rhythm and strung together into attitudes and moods. Initially he emphasizes how varying rhythms and logics of succession generate different feelings:

> That train of thoughts and ideas which is continually passing through the mind does not always move on with the same pace, if I may say so, or with the same order and connection. When we are gay and cheerful, its motion is brisker and more lively, our thoughts

> succeed one another more rapidly. . . . It is quite otherwise when we are melancholy and desponding; we then frequently find ourselves haunted, as it were, by some thought which we would gladly chase away. . . . A slow succession of resembling or closely connected thoughts is the characteristic of this disposition of mind. (162)

The succession of ideas, through its rhythm, supports not only our sense of the expanse of time, but also our sense of its quality because its pace and logic generate feeling. Brisk succession is allied with cheerfulness; slow succession conveys despondence. Smith contends that such varying qualities of duration can arise from the different paces and logics of the succession of sounds, just as much as from those of ideas: "Instrumental Music, by a proper arrangement, by a quicker or slower succession of acute and grave, of resembling and contrasted sounds, can not only accommodate itself to the gay, the sedate, or the melancholy mood; but if the mind is so far vacant as not to be disturbed by any disorderly passion, it can, at least for the moment, and to a certain degree, produce every possible modification of each of those moods or dispositions" (163). Smith makes a strong claim for music's effects: the succession of sounds, in its variations of both sound and pace, not only represents the feelings that accompany such varying succession of ideas, but also produces such feelings in the consciousness of its listener. And he repeats a little later, "Music can, by a sort of incantation, sooth and charm us into some degree of that particular mood or disposition which accords with its own character and temper" (163).

And yet Smith stresses that music doesn't only regulate listeners' feelings by virtue of its shaping the qualities of their durational experience. Much like Addison's methodical essay, music regulates the more general activity of associationist consciousness.

> By the sweetness of its [music's] sounds it awakens agreeably, and calls upon the attention; by their connection and affinity it naturally detains that attention, which follows easily a series of agreeable sounds, which have all a certain relation both to a common, fundamental, or leading note, called the key note; and to a certain succession or combination of notes, called the song or composition. By means of this relation each foregoing sound seems to introduce, and as it were prepare the mind for the following: by its rhythmus, by its

> time and measure, it disposes that succession of sounds into a certain arrangement, which renders the whole more easy to be comprehended and remembered. Time and measure are to instrumental Music what order and method are to discourse; they break it into proper parts and divisions, by which we are enabled both to remember better what is gone before, and frequently to foresee somewhat of what is to come after: we frequently foresee the return of a period which we know must correspond to another which we remember to have gone before; and, according to the saying of an ancient philosopher and musician, the enjoyment of Music arises partly from memory and partly from foresight. (171)

For Smith, music's forte isn't to occupy our duration with its feelings—although Smith has shown that it does this quite well—but to awaken the attention and prompt memory and foresight. At stake are not specific thoughts and contents—representing a melancholy person or telling a sentimental story—but the activity of thinking as an associationist composition.[23] Neil de Marchi points out that despite the title of the essay and the insistent ranking of the effectiveness of different media in making copies, for Smith "the richness of the arts as a source of pleasure cannot possibly be exhausted when we have identified merely the ways and degree to which they imitate."[24] To this assessment I would add Smith's specifying affinities between minds and art that are not, strictly speaking, imitative. Music, unlike some conceptions of figurative sculpture and painting or of realist character, offers only poor copies of the substantive particularities of individuals while superbly instantiating the form of how human minds work by the logic of associationism. Music and minds share the form of temporal composition, crafting wholes out of discrete and disparate moments by activating memory and foresight.

Memory and anticipation have been widely recognized as a key motif in Radcliffe's fiction, and I want now to examine closely how she features this motif and how it is connected to her flat characterization. In flagging recall and anticipation as an insistent yet sparse activity—without the clutter of individuating detail—Radcliffe highlights pleasures of methodical composition that minds and art share in the associationist tradition. Radcliffe's characterization, thus, relies less on a verisimilitude of a subjectivity densely packed with detailed contents and more on a mimesis of the motions of the mind as it integrates past recollection, present perception, and future anticipation.

Radcliffe uses some variant of the words *recollecting, recalling, remembering*, or *memory* approximately 340 times within the six hundred pages of *Udolpho*, in addition to discursive descriptions of processes of recollection that take up much of the space between one literal invocation of remembrance and the next. Scholars tend to examine this motif according to what it might tell us about distinguishing the factual from the fictional. In their repeated demonstrations of how memory and anticipation "clutter the tablet of experience" (applying Frances Ferguson's description of what she takes to be Burke's concern), as well as in their insistence on the supernatural explained, Radcliffe's novels feature worries about our ability to distinguish between what we perceive and what we imagine.[25] Thus, for example, Emily is slow to deter a trespasser in her bedroom because "certain remembrances" and a "late alarming suspicion" (247) absorb her in thought impeding an effective response to the actual noise that's there, in contrast to her dog who, far less capable of imaginings, easily does so. Yet, as Terry Castle argues, while *Udolpho* declares its allegiance to factuality, it is just as much committed to what Castle calls "the supernaturalization of everyday life."[26] Memories of home and loved ones are every bit as substantial—if not more—than the actual presence of these places and persons. And, as Adela Pinch argues, though this novel insists on significant difference between imagination and perception, the factual itself often turns out to be an imaginary product. The veiled object that notoriously terrorizes Emily, leading her to suspect murders and dead bodies, turns out to be not a real corpse but a "waxen image" (622).[27] Thus a sustained analysis of the relation between the factual and imaginary in Radcliffe leads to a dizzying regress.

Radcliffe puts such dizzying regress to rest by recasting at least one kind of mental activity and the representations it yields not as fiction, but as the very compositional activity that constitutes consciousness. While mental images may be distracting, the ability to integrate these into continuous selves turns out to be the crucial test of her characters. We can best see this if we notice the high incidence of "recollecting" and its variants to mean not only memory, but also self-composure. Sometimes the narrator conflates the two meanings, with the self being collected into action by a mechanical memory prompt: "At the sound of his voice, Emily turned her eyes, and a gleam of recollection seemed to shoot athwart her mind, for she immediately rose from her seat, and moved slowly to a remote part of the room" (332). Here recollecting means both recognizing a familiar voice and moving away from it in self-defense. Often Radcliffe privileges the second meaning—self-composure—but has it arise from a dispersal specified temporally: "Having thus continued to converse, without perceiving the flight of time, Valancourt, at length, seemed to recollect himself" (105). And just as often the two kinds of recollection combine to alleviate a distracted state of mind: "Returning reason soon overcame the dreadful, but pitiable attack of imagination, and she turned to the papers, though still with so little recollection, that her eyes involuntarily settled on the writing of some loose sheets, which lay open; and she was unconscious, that she was transgressing her father's strict injunction, till a sentence of dreadful import awakened her attention and her memory together" (99). This last example specifies the adequate state of mind Emily ought to achieve—the awakening at the end of the passage, for which the initial act of recollection falls short—by recasting a competition between reason and imagination into new terms: the conflict of reason and imagination is resolved by the collaboration of memory and attention.

At stake in Radcliffe's "recollection," thus, is not so much "emotion recollected in tranquility," for such recollection *in* tranquility requires temporal and spatial distance, as Wordsworth makes clear. Even when such distance can only be minimally achieved, Wordsworth emphasizes its importance, as in his return to Paris a few weeks after the September massacres, whereby in his room at night "high and lonely, near the roof," he recollects not "wholly without pleasure" the horrors of the square he crossed a few hours earlier, at the time too close to the scene to contemplate its horror.[28] At stake for Radcliffe, by contrast, is emotion recollected *for* tranquility—recollection on the move, akin to the judgment on the move I considered in Richardson in Chapter 2. Radcliffe shares with Richardson a commitment to exploring imaginative projection in time as the experience of the moment, which is both laboriously intense and recruits self-command. But while Richardson is concerned with the ethical dimensions of such a temporally forward-reaching moment, Radcliffe focuses on its therapeutic function and on its aesthetic sources. Radcliffe often associates her heroine's ability to achieve composure through memory with music—both in the thematic sense of having music trigger the action of composure on the move and in the formal sense of allowing her heroine's mind to do the kind of remembering and anticipating that music is charged with in Smith's analysis.

Take, for example, a scene where Emily regains her composure at the castle of Udolpho: "looking on the shadowy scene," she notices how the planets "moved in their destined course" (310), and as Emily "endeavoured to recollect her thoughts, and to reason herself into composure" (310), she recalls her near and distant past. The narration minimally specifies the contents of these memories, telling us that she was "tenderly educated, so tenderly loved," and that she had recently been thrust into "strange and mournful events." Instead of particularizing details, narration emphasizes recurrence of stock events: she sees in the present the planets she once saw with her father, she remembers music she once heard and then hears music presently outside her window. She weeps and then weeps again. The sheer repetition of the syllable "re" in the two paragraphs that precede Emily's success at recollecting herself—some fourteen times through such words as *refreshed, remembered, reflections, retrospect, remote, reason, recollection*—delivers Radcliffe's point most forcefully: at stake is connecting the past to the present, an endeavor most effectively achieved through an active noticing of general resemblances. Then Emily hears music. It's been "very long, since she had listened to any thing like melody," the narrator tells us. And as melody instantiates composition, so Emily's "mind was somewhat more composed" (310).[29] The tune helps the heroine recollect herself as well as prompts her to form hypotheses about the identity of the previous owner of Udolpho—Signora Laurentini who, we learn by the end of the novel, has transformed into Sister Agnes and haunts the landscape by playing mysterious music as though this activity could bridge the gaps in her biography and her soul. But whereas Emily's musically inspired efforts to compose herself as a continuous system succeed, the original owner of Udolpho spectacularly fails. The function of Laurentini/Agnes within *Udolpho*'s character system, to use Alex Woloch's conceptualization, is to raise the value of Emily's activities of recollection: Laurentini/Agnes must fail—haunting the landscape for years and thus turning herself into a ghostly subjectivity—in order for us to grasp what's at stake in Emily's success.[30]

From an ethical perspective, the disparity between Laurentini and Agnes represents the effects of actions on their sources—the distracted Agnes manifesting the burdened conscience of the murderous Laurentini. Within this logic, action is the principle of characters' unity—a character may assume many names and disguises, but it is established and maintained by an unbreakable chain between actions and their effects. This is the Aristotelian subordination of character to action and a juridical subordination of intentions to consequences, which, as we saw in Chapter 2, Sandra Macpherson discovers in *Clarissa*. From such a juridical-ethical vantage point, continuity is given, and the burden of plot and court is to find evidence for it.[31] Radcliffe takes this position in *The Italian* (1797), a conceptual sequel to *Udolpho*. In this novel the endurance of identity is a given while the core problem becomes finding evidence for such consistency despite disguises (new names, new professions, new assumed sensibilities) and other obstructions to perception (not only night, but also dark dungeons, blindfolds, masks, and veils). In *The Italian* voice validates continuity despite all these distractions; clothes and faces may change, but the sound of a voice endures, as evidenced by the numerous scenes in which voice betrays, or alternatively liberates, a character from his or her disguise. Moreover, this novel pointedly literalizes the jurisprudential metaphor of "a hearing," staging its court scenes as occasions where characters can hear but not see each other and dramatizing justice as that special "hearing" in which the hero—Vivaldi—is finally listened to by the court. Radcliffe thus suggests that sound always conveys compositional form; well before sound becomes music, it holds the code that promotes, even guarantees, the sequencing of elements into cohesive wholes.

But in *Udolpho* continuity itself needs to be established—it is *the* problem, rather than an assumption, for which various cases exemplify better and worse solutions. Agnes / Laurentini instantiates one such case: though proved to have a single referent by the end of the novel, she cannot but continue to be understood and designated as two. The beautiful, passionate, and dreadfully focused Laurentini has very little in common with the aged, repentant, and distracted Agnes. Here also Radcliffe displaces the problem of distinguishing imaginary from real experience by turning to the difficulty of organizing experience into cohesive units. The music haunting Languedoc is made no less troubling by the discovery in the last volume that its source is Sister Agnes rather than a ghost, because the problem that Radcliffe wants to register as distressing is Signora Laurentini having transformed irreparably into Sister Agnes.

Establishing a continuous, composed identity underwrites Radcliffe's variation on the gothic's notorious motif of doubles. Gothic fiction's horror often rides on characters being too much like one another—Theodore too much like Alfonso, Isabella too much like Matilda; Caleb too much like Falkland, who turns out to be too much like Tyrrell; Walton too much like Frankenstein, who himself is too much like his monster. Yet even as doubles heavily populate *Udolpho*, such similarities trouble its plot less than do the differences within supposedly same characters as they endure in time. Emily might briefly confuse Du Pont and Valancourt, thinking that the soldier imprisoned in Udolpho is the latter while in fact he is the former. But she never acts according to such confusions—Emily would never substitute Du Pont for Valancourt, although Du Pont becomes her savior and although the Count de Villefort, who now acts as her surrogate father, urges her to marry him instead of Valancourt. While Du Pont occupies Valancourt's position as admirer, savior, and parental favorite, Radcliffe does not allow him to significantly threaten his antecedent. Similarly, while Villefort steps in to compensate for St. Aubert's absence, Radcliffe never indicates any anxiety about father figure usurping father. But even if Du Pont does not threaten Valancourt, the latter's identity is not secure. Valancourt can be imagined or actually be present yet have nearly identical effects on Emily—giving her hope, endowing her with fortitude—while the much more troubling scenes are those in which Emily confronts the question of whether Valancourt remains himself.

Throughout their long and anxious separation in the middle sections of *Udolpho,* Emily sees forms and hears voices that she momentarily mistakes for Valancourt. Yet these mistaken presences hardly breed any plot complications. By contrast, when the lovers finally meet after enduring many longings, Emily startlingly asks Valancourt to "prove . . . you are conscious of being the same estimable Valancourt—whom I once loved" (484). Valancourt, no less alarmingly, cannot confirm that he continues to be himself; he replies, "No—I am not the same!" thus condemning the lovers to more time of anxious separation. Only when Valancourt can be determined as a continuous entity across the narrative—even if by Emily's consciousness more than by his own—can the plot reach its resolution. And Radcliffe stretches this endeavor over the entire fourth and last volume.

Radcliffe deploys in this case what Michael McKeon identifies as the romance's principle of completeness through human connectedness, casting the endurance of love as both the end of her plot and the guardian of the endurance of self. When Valancourt replies, "No—I am not the same!" he also adds, "I am lost—I am no longer worthy of you" (484). Valancourt indicates that his sense of self is constituted not through society as system, which requires continuous selves in order to hold people accountable for their actions, but through the affirmation of an intimate other (with a decidedly small "o") whose role is to reiterate this affirmation in time. But in line with the sensibility chronotope, Radcliffe also crucially temporalizes such completeness founded on human connectedness, casting it as the necessity of recursive affirmations rather than as a singular event of closure. As Valancourt knowingly tells Emily early in their courtship, "Such is the inconsistency of real love, that it is always awake to suspicion, however unreasonable; always requiring new assurances from the object of its interest, and thus it is, that I always feel revived, as by a new conviction, when your words tell me I am dear to you; and, wanting these, I relapse into doubt, and too often into despondency" (146). But in an era in which the substantiality of others becomes increasingly doubtful, as Castle interprets the popularity of ghost stories at the turn of the nineteenth century, one cannot rely on such recursive affirmations of love in presence to guarantee the continuity of self.[32] Valancourt is lucky to have Emily, but Emily can take up the role of confirming his endurance precisely because of her recursive remembering of Valancourt and her confirmation of their love over and over again, even when he cannot hear her. If Valancourt cannot recall Emily in her absence, Emily's incessant memories suffice to confer composure on both of them. But Emily's love does not make her or Valancourt individuals; these characters are partial doubles of one another, and of many other characters in the novel. Rather, Emily's incessant recollections make her a figure of human consciousness—an associationist composite that shares with music the pleasure and form of composing durational coherence out of succession.

## Recollection and the Romance of Character

Radcliffe's recurring use of "recollect" to mean both memory and self-composure encapsulates one key way in which she revises Walpole's gothic romance. If Walpole's inaugural instance of British gothic fiction has anything to do with his famous antiquarian predilection for collecting, then Radcliffe recasts collecting as recollecting with all that the added prefix entails. Walpole repudiates temporal transcendence by staging the tragedy of *The Castle of Otranto* as the visitation of fathers' inequities on children and grandchildren—a collapse of past, present, and future that pronounces the eclipsing of time as moral horror. The historical conceit of his fiction functions as an antidote to such horrors of timelessness, for though providence can't tell the difference between one generation and the next as it metes out punishment, in the first preface Walpole published with his romance, he takes pains to distinguish the time of printing, from the time of composing, from the time of represented events, as he makes the case for his story's significance.[33] Yet these historicist commitments are predicated on fiction, which makes them seem questionable to many scholars who study Walpole's antiquarian ambitions in *Otranto* as well as in his collections displayed at Strawberry Hill. In an era as concerned with historical consciousness as the eighteenth century, Walpole's privileging of rhetoric and style over factual authenticity—his equivocating the difference between actual relics from the past and ones crafted in the present to resemble those from the past—is especially troubling.

Ruth Mack persuasively argues that the flimsiness of Walpole's historical references—like the thinness of the paper turrets at Strawberry Hill—articulates a skeptical position about our ability to achieve historical consciousness. Working at the cusp of a transition from a textual historiography to one in which objects gain evidentiary status, Walpole challenges the belief that objects can give us an experience of the past as though we might have been present in it.[34] Radcliffe neither aims to achieve nor repudiates representations that wrestle with the question of historical consciousness. Though *The Mysteries of Udolpho* begins with announcing its setting in the year 1584, there's little in the rest of the book to mark calendrical time, and as many readers notice, its characters instantiate the culture of eighteenth-century England much more than that of sixteenth-century France. But if Walpole addresses the urgent problem of temporal consciousness by directing his attention to relics and historicism—external and normative indexes of time passing—and flagging the difficulties of this solution, Radcliffe turns to consciousness as such. If Walpole engages time passing with the antiquarian's compiling of familial, religious, and national histories through objects, then Radcliffe considers it as a person's taking stock of his experience through thought. And finally, if Walpole approaches time as the historicity that inheres in objects (whether naturally or artificially), then Radcliffe considers temporality as a cognitive process.

This is not to argue that Radcliffe psychologizes Walpole's approach to history. Walpole conceives of his gothic as fusion of ancient and modern romance—modern romance standing for what we today call the novel—detailing this fusion as the psychologization of character. Though in his first preface to *Otranto* Walpole makes the case for the "realness" of his tale by way of sham claims to historicity, in the second preface he openly engages the question of the reality of fiction, explaining that he has applied "the boundless realms of invention" to devise "more interesting situations," while devising "mortal agents . . . according to the rules of probability; in short, to make them think, speak, and act, as it might be supposed mere men and women would do in extraordinary positions" (21). But while Walpole's psychologization is predominantly historicist and spatial, Radcliffe's is sensible and temporal. Walpole thinks of duration as discrete segments among which one can at best leap through fictional means; Radcliffe, by contrast, understands it as a dynamic process whose very processual nature one must actively compose. And if Walpole approaches time through the content carried by its segments—examples of ancient versus modern fiction, examples of gothic versus neoclassical architecture, examples of older versus younger generations—Radcliffe conceives of it through form—less the content to which one returns (though obviously memories featured in the novel have some minimal degree of content) than the compositional effort that enables return. Both these temporal conceptions rely on mental activity to supplement perception, but while collecting entails recognizing the less than factual status of the past—Walpole's point about our leaps of faith in seeking authenticity in objects—recollecting indicates that relations among discrete segments are constructed and thus inhere in the mental form of composing rather than in the content of any given moment. A final way to characterize the contrast between Walpole and Radcliffe is to say that Walpole adheres to a spatial model of the mind as a collection, as Sean Silver delineates it, while Radcliffe takes up its temporalized articulations—the ones I've been tracking in this book as participating in the sensibility chronotope.[35]

Radcliffe differentiates the gothic and sensible casts of *Udolpho* along the fault line of durational composition. The monstrous—inhuman—characters of the novel fail to endure; they either splinter into two—as in the case of Laurentini/Agnes—or disappear into none. Montoni, the arch villain of the middle sections of *Udolpho*, turns out to have very little effect on the plot's outcomes; Emily and Valancourt unite, Laurentini is alive and kicking, and even Madame Montoni successfully refuses signing her property over to her husband and dies less at his hand than by her own determination. By the last volume of the novel Montoni all but disappears, his demise not "attracting curiosity, or even . . . obtaining a place in any of the published records of that time" (492), and his death imparted in a mere sentence (536). The novel's sensible characters, by contrast, endure by the clearly flagged effort of composing, whether of another's love or of their own recollecting, coming to seem human not through nuanced and complex subjectivities, but through recurring references to the kind of attention that promotes associationist activity.

Scenes of recollecting that produce consciousness as collecting across a temporal axis, along with scenes of nocturnal meditation, as well as *Udolpho*'s distinction between gothic and sensible characters, all indicate that Radcliffe understands the new romance dimension of her fiction—its psychological mimesis—to reference the human mind in its fundamental, typical associationist capacities, rather than to reference individuals in their content-laden particularities.[36] But this model of characterization does not make Radcliffe an outlier—more of an "old romance" writer than other contemporary and subsequent practitioners of psychological realism. Again, Jane Austen's *Sense and Sensibility* can serve as an example. That Austen, in *Sense and Sensibility* especially, typifies her characters, no less than individualizes them, need not be established. Deidre Lynch demonstrates this, appealing to the impersonality of markets to argue that "Austen's novels position interiority at a relay point that articulates the personal with the mass-produced."[37] But I want to highlight Austen's concerns with questions of endurance and recollection in her typified characters. More than any of her other novels, *Sense and Sensibility* tackles the problem of temporal instability, not only as a matter of plot (as we've seen in Chapter 2) but also as a matter of characterological consistency. The novel takes up the model of active mental composing through stock resemblances as a viable technique for realist characterization, although for Austen it is only one among several models.

Elinor's character engages time primarily as a medium of disclosure, and fittingly her favorite art form is drawing. Like the landscapes she draws, all is there in her character from the start, time only serving to discover and improve. But for Marianne duration constitutes identity, an identity that is most impressive as it survives the change from being a skeptic of the possibility of change to being living proof of its inevitability. Having fallen in love with the type of rake who is also a gothic villain—not only attuned to his present moment of pleasure, but also a cunning economic opportunist who abides neither by juridical, nor sentimental, nor aesthetic models of consistency—Marianne's initial commitment to static essences—single loves and opinions to last a whole lifetime—must give way to a more dynamic conceptualization of endurance. And as befitting a character for whom temporal consistency is the core problem, Marianne loves music. But although playing the pianoforte offers ample opportunities to exercise the kind of compositional attention that makes melody valuable for Smith and for Radcliffe, it takes a substantial threat to Marianne's life for her to apply music's lessons to her own endurance. Only after surviving a life-threatening fever she take up cognitive self-composition: "My illness has made me think—It has given me leisure and calmness for serious recollection" (293), she tells Elinor.[38]

Austen also uses sensible typification to craft Brandon's character—for Brandon's steadfast love vouchsafes Marianne's consistency similarly to Emily's recurring affirmations of love serving to compose Valancourt. Austen models Brandon most directly on Radcliffe's sensible recollection; if Willoughby is a version of Montoni, then Brandon is an Emilied Valancourt. Like Emily, Brandon incessantly recollects: the archetype of the steadfast lover whose commitment to Marianne does not flinch through her attachment to, and disappointment in, Willoughby. Moreover, Brandon falls in love with Marianne instantly only because she resembles his first love, Eliza, underlining the deindividualized constitution of his memory. But Austen also emphasizes that a conscious effort of recollection, in both meanings that are key for Radcliffe—remembrance and composure—crucially supports Brandon's consistency. When Brandon relates his biography to Elinor, he uses the vocabulary of self-composure. Right before he begins his narrative, "He stopt a moment for recollection" (173); midway through recounting Eliza's story he exclaims, "even now the recollection of what I suffered—" (174); after which he requires "a few minutes more of silent exertion" before he can "proceed with composure" (174–75). And the confession concludes when "recollecting, soon afterwards, that he was probably dividing Elinor from her sister, he [Brandon] put an end to his visit" (178). Austen thus suggests that this "sensible man" (45), as Elinor identifies Brandon earlier in the novel, is coherent as well as typical by virtue of relentlessly taking up the effort of recollection.

Much of what I have been observing about Marianne and Brandon has already been noticed by Nicholas Dames, who argues that "an ethic of narratological health governs the text's [*Sense and Sensibility*'s] moments of remembrance" (50).[39] But according to Dames, Austen inaugurates the logic of normative amnesia—narratological health construed as the kind of nostalgia "which stiffens the self against the weight of mnemonic diffusion that later psychologists and novelists would find so compelling" (5). And yet one has to presume the truth of the later psychologists—that "mnemonic diffusion" threatens an already established selfhood—in order to explain consolidation as amnesia. Given *Sense and Sensibility*'s firm anchor in the sensibility chronotope, and given its characters' explicit echoing of Radcliffe's sensible gothic, it seems to me more likely that Austen takes up "narratological health" as the primary associationist activity that labors to establish the very continuity of consciousness, rather than presuming continuity only to erase some of its content. From the perspective of eighteenth-century associationism, before one can strategically forget—become amnesiac in Dames's sense—one has to be able to conceive of the self as one. For the associationists, diffusion is atemporal—rather than, to begin with, a disease of memory—and memory and anticipation serve to produce a coherent temporalized consciousness from this atemporal diffusion.[40]

But if Dames's study of amnesia and novelistic characterization misses Austen's beholdenness to sensibility's associationism, his study of what he calls physiological novelists and critics seems to me helpful in recognizing what might be sensibility's afterlife in Victorian characterization. Dames considers a group of writers who focus on the temporal dimensions of novels, use music as their compositional model, and promote a notion of readerly pleasure that arises from the typicality, rather than the individuality, of experience. "Physiological concepts taught the critics who adapted them for literary analysis that form, particularly novelistic form, must be defined temporally, as a rhythm or time signature, rather than a synchronic structure. In musicological terms, physiological novel theory was interested in horizontal, or rhythmic-melodic, analysis rather than vertical, or harmonic analysis," Dames explains.[41] These physiological novelists and critics rarely refer to writings prior to the 1850s, according to Dames. But we might hypothesize their eighteenth-century prehistory: the sensationist and associationist models of durational aesthetics, developed in philosophical and musicological treatises and in Radcliffe's—as well as in Sterne's and Richardson's—sensible novels.

# Coda

## The End of Human Time?

I HAVE CONCLUDED each of the preceding three chapters with discussions of Jane Austen's *Sense and Sensibility* aiming to open up a wider historical horizon for the sensibility chronotope. My short readings of Austen's novel have focused on dimensions of compositional crafting that nineteenth-century English novelists adopted from eighteenth-century precedents and adapted for their purposes—the plot of the second take as it revises sensibility's durational ethic, rhythmic narration and its conduciveness to sustainable sympathy, and the musical model of character in its approach to the problem of the continuity of identity. These discussions, as well as my analyses throughout the book, have underlined the inextricability of ethical, aesthetical, and formal problems. From the perspective of the sensibility chronotope, what we do, how we feel, and who we are, are all questions that have everything to do with the temporality of our existence, and the temporality of our existence has everything to do with the way our experiences are composed into sensible patterns. I want to conclude this book by expanding even further the historical horizon of the sensibility chronotope, in its approach to time as both an experience and an ethic that is constituted by sensual and compositional arrangements.

Recall that the point of departure for John Locke's empiricist account of duration was the alignment of temporality with existence—I think therefore I endure. But while my explorations have focused on the qualitative potentials that this alignment supports, I only touched in passing on how this premise highlights the finiteness of human time and, by extension, associates the capacity to experience qualities with the ineluctability of an end. Such emphasis on finitude underwrites many gothic renderings of the eclipse of time as horror, as well as more recent alarming discussions of our 24/7 culture. Charles

Maturin's *Melmoth the Wanderer* (1820) instantiates the association of ruthlessness with immortality; Mary Shelley's "Mortal Immortal" (1833) exemplifies the alignment of endlessness with despair. Both suggest that an existence without the consciousness of an end is doomed to incessant counting, which drains sensible qualities from durational experience. The image of a recursively ticking clock comes to be exemplary of a horrific coupling of monotony and infinity, while the image of a human being who changes in time and as time comes to instantiate a reassuring pairing of variability and finiteness. Such gothic renderings anticipate Jonathan Crary's recent articulation of this contrast as an imminently disastrous contest. As Crary puts it, "the imposition of a machinic model of duration and efficiency onto the human body," and "the generalized inscription of human life into duration without breaks, defined by a principle of continuous functioning," effects "a static redundancy that disavows its relation to the rhythmic and periodic textures of human life."[1] This 24/7 chronotope, according to Crary, traps us in a struggle for domination, which machines are perilously close to winning. But there may be other ways to understand the relation between human duration and machinic time. As I have suggested in passing in the previous chapters, the sensibility chronotope harnesses the recursivity of fiction and print—their mechanical and hence iterable capacities—for its own purposes of communicating qualitative experience. Similarly, the recursivity of clocks can be used as a means for composing sensible duration, rather than as a disciplinary golem that comes to rule us. Christian Marclay has recently made such a case with his montage installation *The Clock*.

With a running time of twenty-four hours, Marclay's *The Clock* (2010) assembles clips from movies and television, each somehow indicating the hour and minute, and together arranged to correspond perfectly with the actual chronometric time of its audience. As an audience member, you walk into the gallery, often after waiting in line outside for a while. You settle down on one of the comfortable sofas—props dictated by Marclay's installation directions, designed to maximize impact by shaping the body positions of its viewers—and begin to see and hear the time relentlessly. Close-ups of wrist watches, mantel pieces, pocket pieces, alarm clocks, train station clocks, school and bank clocks, and Big Ben (of course); familiar screen faces consulting their watches, discussing this time or that, and occasionally contemplating time in general; soundtracks of ticks, tocks, tolls, chimes, and bangs. Immersed in these surroundings, you come to think that the transnational and transhistorical star of cinema and television had been none other than the clock. The leading leisure genres of our contemporary modernity relentlessly remind us of mechanical time-keeping, which seems to aspire to occupy every single minute of day and night.

On a cursory look, then, Marclay is in the business of hammering in our resignation to mechanical reproduction, with its negation of qualitative duration and its privileging of homogenous empty time. Indeed, if we presume that qualitative experience ought to somehow rely on singular particularity, then seeing all these simulacra of clocks cannot but make us lament the monotony of our technological predicament. And yet the effects of this installation turn out to be very different, as many testimonials indicate. For even as chronometry is inescapable in "The Clock," and even as screen time corresponds perfectly with the time on one's own watch (or smartphone), people lose their count of time for the duration of their viewing. "I left Christian Marclay's 'The Clock' about twelve minutes after 'High Noon.' I'd been there since about 10:30 a.m., thinking that I'd stay for about twenty minutes and get the feel of the thing; any longer, and the minutes might drag. But I was wrong. Time flew—it hypnotized," one critic testifies in a review entitled "Is 'The Clock' Worth the Time?"[2] "After watching 'The Clock' from around 7:30 p.m. last Friday to past midnight, I dragged myself away, despite the desire to stay," another confesses.[3] "For me, the weirdest effect of The Clock is that the time references became fictional—I stopped noticing that they were telling me exactly what the time actually was. They became a series of numbers which ordered the mosaic of moods and moments. And then, slowly but surely, I stopped noticing the time entirely. I just drank it in, just accepted the juxtapositions," another explains.[4] All of these testimonials chime well with my own experience. And my partner's. And our friends'. And the strangers we chatted with while queuing to get in a few times in New York and a few more in Minneapolis.[5] Marclay immerses his viewers in chronometry only to produce a radical transformation in their sense of duration: initially hyperaware of time's measure, the audience of "The Clock" quickly allows itself to lose count.

But this is an arrest that by no means produces temporal transcendence. Rather, *The Clock* highlights the conditions wherein counting time becomes conducive for experiencing qualities of duration, wherein chronometry comes to serve what I have identified in this book as the sensibility chronotope. If the conspicuous compositional principle of *The Clock* is piecing together

1,440 minutes of clock time, then Marclay's montage enables it to become a masterpiece of qualitative duration—giving his viewers an especially intense and thoroughgoing feel of moments. Often, as we get close to the hour pace escalates. At around 10 a.m., as late sleepers awake, quick rhythms alternate between alarm and indulgence. At around 4 p.m., as workers look forward to ending the workday and children to ending the school day, we sense rhythms of joy. As we approach high noon, as well as midnight, duration becomes at once unbearably suspenseful and remarkably slow. The wee hours of night as well as siesta time—between 2:00 and 3:00 a.m. and p.m.—invoke, by contrast, an atrophying slowness. Marclay's masterful editing facilitates such transmogrifying of measurement into qualities. He seams together soundtracks to accentuate rhythm—whether in ticking clocks, running taps, chiming bells, strings of language or of music. He carefully attends to lighting, color, and surfaces. He juxtaposes camera angles to raise anticipation. He returns us to the same scenes, movies, or actors several times within an hour. And it is the combination of these editing techniques that compels our minds to connect, to remember, to anticipate even more than any dramatic plot line might have done. In short, *The Clock* guides the succession of its audience's ideas—to invoke Locke's primary definition of duration one last time—controlling the pace, mood, and directionality of our thoughts, and doing so less by means of the succession of causal action, and more by patterns of shape and sound.

"Classic film editing doesn't enchain time. Instead, it exchanges one moment's spatial visualization for the next in a framed plastic series," Garrett Stewart explains. He continues: "Montage in cinema actually *produces* (rather than captures) narrative time out of spatialized event, where the graphic series is read not as continuous duration but as an *articulated* temporal sequence. Montage doesn't in this sense take time; it makes it."[6] Stewart's definition of montage is especially useful for my purposes for its highlighting the temporal ambitions of the technique, even as he might be overemphasizing its spatiality. For one virtue of Marclay's editing is to rely heavily on the work done by soundtracks and thus to help us notice how montage exploits sonorous imaging and not just spatial visualization. Stewart himself knows this well insofar as he studies montage's prehistory in novelistic techniques of Victorian high realism, attending not only to visual images but also to minute linguistic levels of sonorous sequencing—as we have seen in Chapter 3.[7] But, as James Chandler more recently argues, the prehistory of montage might be backdated even further to the technical innovations of eighteenth-century sensibility. Chandler gives us a compelling new perspective on what he calls "the sentimental disposition," demonstrating its development in eighteenth-century sensibility—literature and moral philosophy—and its continued influence up to the narrative system of classical Hollywood cinema. In Chandler's account, sensibility develops a form that highlights facial expressions in sequences of alternating points of view, and this constitutes the sentimental effect of novels and, subsequently, of movies.[8] But if, for Chandler, juxtapositions of facial close-ups elicit sentiment, then in Marclay's work juxtapositions of visual and sound images of clocks, and not just of human faces, do the job, and do so to mobilize feelings of and about time, rather than those of and about space.

The preceding pages have not made the case for an alternate prehistory of montage. And yet we find in Marclay's *The Clock* many techniques and effects that we've seen developed and mobilized in novels of sensibility, suggesting that while the technological tools at Marclay's disposal are new, the logic of form informing his work has been with us for a while. And even more important for my purposes here, Marclay's montage reminds us that the recursivity and conspicuousness of clocks can be used to make a very different point from the one about the triumph of machinic monotony. Once we recognize the logic of the sensibility chronotope—that qualitative duration is, to begin with, both sensationist and compositional, which is to say that iterability counts as one of its constitutive dimensions—then chronometry, just like the technologies of print and of fiction, can become none other than a prop for mediating a thoroughly human time.

# *Notes*

## INTRODUCTION

1. Daniel Defoe, *Robinson Crusoe*, ed. Michael Shinagel, 2nd ed. (New York: W. W. Norton, 1994).

2. "From the 1st of *October*, to the 24th. All these Days entirely spent in many several Voyages to get all I could out of the Ship, which I brought on Shore, every Tide of Flood, upon Rafts" (52), he tells us. "I was no less Time than from the 3d of *January* to the 14th of *April*, working, finishing, and perfecting this Wall" (56), he explains of another activity. "I built me a little kind of a Bower . . . And this Work took me up [from July 19th] to the Beginning of *August*" (75), he explains of yet another.

3. More evidence for this temporal shift in *Robinson Crusoe* can be found in the novel's insistent characterization of Friday through his pace, and in its staging of the encounter between the two men as a contest and a relationship between their personal times. For a fuller case for such a qualitative turn in *Robinson Crusoe*, see my "Time, Duration, and Defoe's Novels," *Partial Answers* 6, no. 1 (2008): 33–56. Also see Paul Alkon's *Defoe and Fictional Time* (Athens: University of Georgia Press, 1979), which argues that Defoe anticipates modernist temporal phenomenologies as well as draws on seventeenth- and early eighteenth-century narrative models.

4. E. P. Thompson discusses the shift from task-oriented labor to industrial labor and argues that it entails a departure from irregular rhythms and personal, human-centered pace-makers and the adoption of standardized rhythms and mechanical pace-setting. See "Time, Work-discipline and Industrial Capitalism" in *Customs in Common: Studies in Traditional Popular Culture* (London: Merlin Press, 1991). Michel Foucault demonstrates how techniques for ever-growing partitioning and micromanagement of time and space have been conducive for producing docile bodies. See "Docile Bodies" in *Discipline and Punish: The Birth of the Prison*, trans. Alan Sheridan (Harmondsworth: Penguin, 1977). And Benedict Anderson underlines a transition from messianic conceptions in which distinctions of past, present, and future collapse, to a secular approach that takes time as a neutral medium in which a group of people can progress together. See *Imagined Communities* (London: Verso, 1983). The modern timekeeper chronotope can be described as "precise, punctual, calculable, standard, bureaucratic, rigid, invariant, finely coordinated, and routine" (xvi), as the sociologist Eviatar Zerubavel puts it; see

*Hidden Rhythms: Schedules and Calendars in Social Life* (Chicago: University of Chicago Press, 1981).

5. John Locke, *An Essay Concerning Human Understanding*, ed. Peter H. Nidditch (Oxford: Oxford University Press, 1975), 396.

6. David Hartley, *Observations on Man, His Frame, His Duty, and His Expectations* (London, 1749), 65.

7. Bishop Berkeley criticizes Locke for insufficiently acknowledging the incongruities between his accounts of our sense of duration and absolute time. Berkeley writes in his *A Treatise Concerning the Principles of Human Knowledge* (Dublin, 1710): "Whenever I attempt to frame a simple idea of *time*, abstracted from the succession of ideas in my mind, which flows uniformly and is participated by all beings, I am lost and embrangled in inextricable difficulties. I have no notion of it at all, only I hear others say it is infinitely divisible, and speak of it in such a manner as leads me to entertain odd thoughts of my existence; since that doctrine lays one under an absolute necessity of thinking, either that he passes away innumerable ages without a thought, or else that he is annihilated every moment of his life, both which seem equally absurd" (Part I, §. 98).

8. My understanding of Edmund Husserl's, Henri Bergson's, and Martin Heidegger's projects relies on Husserl's *On the Phenomenology of the Consciousness of Internal Time (1893–1917),* trans. John Barnett Brough, vol. 4 of *Edmund Husserl, Collected Works* vol. IV, ed. Rudolf Bernet (Dordrecht: Kluwer Academic Publishers, 1991); on Bergson's *Time and Free Will: An Essay on the Immediate Data of Consciousness*, trans. F. L. Pogson (London: Dover, 2001); and on Heidegger's *Being and Time*, trans. John Macquarrie and Edward Robinson (Oxford: Blackwell, 1962). Studying these difficult works over the years, the following secondary sources have been especially helpful: David Couzens Hoy, *The Time of Our Lives: A Critical History of Temporality* (Cambridge, MA: MIT Press, 2009), Philip Turetzky, *Time* (London: Routledge, 1998), *The Routledge Companion to Phenomenology*, ed. Sebastian Luft and Soren Overgaard (London: Routledge, 2012) and in it especially Nicolas De Warren's essay "Time," and William Blattner, "Temporality" in *A Companion to Heidegger*, ed. Hubert L. Dreyfus and Mark A. Wrathall (Malden, MA: Blackwell, 2005).

9. Several scholars of eighteenth-century literature have applied conceptions developed by twentieth-century phenomenology—Bergson's, Husserl's, or Heidegger's—to their readings of novels. With the exception of Alkon's study of Defoe, such applications tend to approach the early literature as either anachronistic anticipations of, or negative counterpoints for, modernism. See A. A. Mendilow, *Time and the Novel* (1952; rpt. New York: Humanities Press, 1972), Benjamin H. Lehman, "On Time, Personality, and the Author" rpt. in *Laurence Sterne: A Collection of Critical Essays*, ed. John Traugott (Englewood: Prentice Hall, 1968), and Jean-Jacques Mayoux, "Variations on the Time-sense in *Tristram Shandy*" in *The Winged Skull*, ed. Arthur H. Cash and Joan M. Stedmond (Kent, Ohio: Kent State University Press, 1971).

10. In *Virtue in Distress: Studies in the Novel of Sentiment from Richardson to Sade* (London: Macmillan, 1974), R. F. Brissenden examines relations between three unstable terms: sentimental, sensible, and experience. Ann Jessie Van Sant tracks relations between

sensibility, sentiment, and delicacy in the first chapter of *Eighteenth-Century Sensibility and the Novel: The Senses in Social Context* (Cambridge: Cambridge University Press, 1993). Jessica Riskin offers the following explanation for the relation between sentiment and sensibility: "sentiment" is defined by Diderot as "an emotional 'movement' in response to a physical sensation" and sensibility as "the first germ of thought" giving rise to "Moral sentiments" by way of "openness of the soul to vivid impressions of a delicate world"; *Science in the Age of Sensibility: The Sentimental Empiricists of the French Enlightenment* (Chicago: University of Chicago Press, 2002), 1–2. James Chandler proposes that the discourse of moral sentiments and the discourse of sensibility have been imbricated through conceptions of emotion that link it with motion; see *An Archaeology of Sympathy: The Sentimental Mode in Literature and Cinema* (Chicago: University of Chicago Press, 2013), chapter 5.

11. For the origins of sensibility in philosophical and physiological studies, see G. S. Rousseau's "Nerves, Spirits, and Fibres: Towards Defining the Origins of Sensibility," in *Studies in the Eighteenth Century, III: Papers Presented at the Third David Nichol Smith Memorial Seminar, Canberra 1973* (Toronto: University of Toronto Press, 1976); see also his "Discourses of the Nerves" in *Literature and Science as Modes of Expression*, ed. Frederick Amrine (Dordrecht: Kluwer Academic Publishers, 1989), 29–60. G. J. Barker-Benfield in *The Culture of Sensibility: Sex and Society in Eighteenth-Century Britain* (Chicago: University of Chicago Press, 1992), offers a discussion of Newton and Locke, as well as an account of sensibility's underwriting of eighteenth-century domesticity and consumerism; also see the excellent introductory essay on sensibility in *An Oxford Companion to the Romantic Age: British Culture, 1776–1832*, gen. ed. Iain McCalman (Oxford: Oxford University Press, 1999), 102–14. John Mullan in his *Sentiment and Sociability: The Language of Feeling in the Eighteenth Century* (Oxford: Clarendon Press, 1988) considers novels in relation to the moral philosophy of Hume and Smith on the one hand and medical treatises on the other. Lynn Festa makes the case for the way in which sentimental rhetoric regulates the circulation of feeling along national borders and imperial hierarchies in *Sentimental Figures of Empire* (Baltimore: Johns Hopkins University Press, 2006). Scott Paul Gordon, in *The Power of the Passive Self in English Literature, 1640–1770* (Cambridge: Cambridge University Press, 2002), discovers sensibility's promotion of class distinctions by way of differentiations in aesthetic attitudes. And Deidre Shauna Lynch in *Loving Literature: A Cultural History* (Chicago: University of Chicago Press, 2015) makes the case for sensibility's supporting a new subjectivity that conjoins feeling and knowing through its elaborations of literary appreciation.

12. Brissenden, Barker-Benfield, Riskin, Chandler, and Lynch all examine sensibility's capacious and long-lasting influence.

13. For the subtlety and complexity of approaches to feeling, see Chandler and Lynch, as well as Hina Nazar, who argues that British sentimentalist philosophers and novelists sought to formulate normative precepts that rely on intersubjective and emotional modes of reasoning. See her *Enlightened Sentiments: Judgment and Autonomy in the Age of Sensibility* (New York: Fordham University Press, 2012).

14. Giorgio Agamben, "Time and History: Critique of the Instant and the Continuum" in *Infancy and History: On the Destruction of Experience*, trans. Liz Heron (London: Verso, 2007), 91. The circuit of temporality and culture is an especially difficult one, overlooked by the more direct applications of modernist phenomenologies to eighteenth-century literature, as well as by more general philosophical perspectives on time.

15. Mikhail Bakhtin, *The Dialogic Imagination: Four Essays*, ed. Michael Holquist, trans. Caryl Emerson and Michael Holquist (Austin: University of Texas Press, 1981).

16. See Chandler, *An Archaeology of Sympathy*, especially chapters 4, 5, 6.

17. Bakhtin identifies five strands of idyll-nostalgia in the modern novel. First, "the influence of the idyll, idyllic time and idyllic matrices on the provincial novel" (228–29); here idyll is interiorized, individualized, nostalgic. Second, "the destruction of the idyll, as in the *Bildungsroman* of Goethe and in novels of the Sternean type" (229), with two strands: one in which idyll is treated nostalgically (233), the other in which its inadequacy for the modern capitalist world is brutally exposed (Flaubert exemplifies this). Third, he notes the "influence on the Sentimental novel of the Rousseauan type" (229) in which interiorization and nostalgia are progressive—a critique of the present and an opening of future positive horizons (231). Fourth is "the family novel and the novel of generations": *Tom Jones* exemplifies one strand, in which the threats of an alien, random world get resolved by the stable love-based world of family; *Clarissa* exemplifies another strand in which "an alien force intrudes into the cozy little world of the family, threatening it with destruction" (232). And finally, the baggiest of all of Bakhtin's classifications is novels "belonging to certain other categories" (229). The sprawl of Bakhtin's analysis may suggest that at stake is more than a single chronotope.

18. Stuart Sherman, *Telling Time: Clocks, Diaries, and English Diurnal Form, 1660–1785* (Chicago: University of Chicago Press, 1996).

19. Such privileging of privacy and isolation is most fully developed in Sherman's readings of Burney's diaries that align the interruptions of their diurnality with the feminine plight of lack of a room of one's own in which to pursue one's writing in solitude (247–68, and esp. 251–52).

20. See Lynch, *Loving Literature*, especially 157–58.

21. Also see Lynch, "Austen Extended/Austen for Everyday Use" where she points out that "the novel's cultural ascendancy was also a function of readers' esteem for the way that this genre is able, as no other, to simulate the feel of duration, the feel that time has when apprehended as a steady state. The novel takes its time and so fills up plenty of ours" (*Imagining Selves: Essays in Honor of Patricia Meyer Spacks*, ed. Rivka Swenson and Elise Lauterbach [Newark: University of Delaware Press, 2008], 238).

22. William Wordsworth, *Lyrical Ballads*, ed. Michael Mason, 2nd ed. (Harlow, UK: Pearson Longman, 2007), 64.

23. M. H. Abrams, *Natural Supernaturalism: Tradition and Revolution in Romantic Literature* (New York: W. W. Norton, 1971).

24. Michael Clune, *Writing Against Time* (Stanford: Stanford University Press, 2013). Clune contrasts the Romantic approach with a Classical one, which he takes to be formalist rather than experiential. The Classical, Clune explains, takes the beautiful form of the

poem to transcend the decay of its individual readers and enables the author to endure as a reputation represented by the poem as a formal expression of the desire for immortality. Life as embodied experience becomes "a sacrifice" (6) for form in this model. The Romantic tradition, by contrast, aims to give readers a technology for a lasting experience. Put in this way, Clune's discussion resonates with the possibility that the Romantics didn't so much aim to defeat time as to reinfuse it with duration. But this is not Clune's argument.

25. Christopher R. Miller, *The Invention of Evening: Perception and Time in Romantic Poetry* (Cambridge: Cambridge University Press, 2006) and Julia S. Carlson, *Romantic Marks and Measures: Wordsworth's Poetry in Fields of Print* (Philadelphia: University of Pennsylvania Press, 2016).

26. Ian Duncan, *Modern Romance and the Transformations of the Novel: The Gothic, Scott, Dickens* (Cambridge: Cambridge University Press, 1992), 21.

27. Elizabeth Deeds Ermarth, *Realism and Consensus in the English Novel: Time, Space, and Narrative* (Princeton: Princeton University Press, 1983).

28. Michael McKeon, *The Origins of the English Novel, 1600–1740* (Baltimore: Johns Hopkins University Press, 1987).

29. For Paul Ricoeur's reading of Aristotle, see chapters 2 and 3 in vol. 1 of *Time and Narrative*, trans. Kathleen McLaughlin and David Pellauer, 3 vols. (Chicago: University of Chicago Press, 1984). For his discussion of narrative structure in fiction, see vol. 2, and for his phenomenology of reading, see vol. 3, chapter 7.

30. Ricoeur notes semiotics' attempts to subordinate "every syntagmatic (and therefore temporal) aspect of narrative to a corresponding paradigmatic (and therefore achronological) aspect" (*Time and Narrative*, 2: 31), as he puts it in his refutation of certain kinds of structuralism. Catherine Gallagher diagnoses a similar problem, explaining that formalist discussions of seemingly opposing commitments all tend to feature "a bias against the very thing under analysis, that is, extended temporal sequence—length—itself" (306), and calls for studies that gauge "the temporal nature of narrative" ("Formalism and Time" in *Reading for Form*, ed. Susan Wolfson and Marshall Brown [Seattle: University of Washington Press, 2006], 327).

31. Elaborating on Mendilow's distinction between "tales of time" and "tales about time," Ricoeur explains that "all fictional narratives are 'tales of time' inasmuch as the structural transformations that affect the situations and characters take time. However only a few are 'tales about time' inasmuch as in them it is the very experience of time that is at stake in these structural transformations" (2: 101).

32. "Ought Sensibility To Be Cherished, or Repressed?" *Monthly Magazine* 9 (October 1796): 706–9. Ricoeur attributes his dismissal of eighteenth-century novels to what he (mistakenly) takes as their naïve realism; he writes: "Since the birth of the novel through the end of its golden age in the nineteenth century, a more urgent problem than that of the art of composition occupied the foreground: the problem of verisimilitude. . . . It was this concern for being true—in the sense of being faithful—to reality, or for equating art and life, that most contributed to covering over the problems of narrative composition" (2: 10–11).

33. Jonathan Lamb's study of Sterne's double principle offers a compelling statement about sensibility's suspicion of conceptual abstractions. Lamb examines the skepticism generated by a doubleness of literal and figurative meaning. Sterne, he explains, is "fascinated by the confusion of the intellectual and the material worlds that comes about when words straddle the border between figurative and literal meanings. Like the transformations it accompanies, this wordplay enacts a repudiation of the thing itself" (*Sterne's Fiction and the Double Principle* [Cambridge: Cambridge University Press, 1989], 30). A number of more recent studies have sought to explain the focus on feeling and embodiment through wider cultural trends. See Van Sant's discussion of relations of vision and touch in sensibility. Also see Paul Goring's *The Rhetoric of Sensibility in Eighteenth-Century Culture* (Cambridge: Cambridge University Press, 2005), which considers novels within the context of a wider elocutionary discourse, arguing that all these have a didactic aim of acculturating readers in specific normative somatic performances.

34. On the episodic and fragmentary form of novels of sensibility, see Van Sant. On the spatial component of multiperspectivalism, see Chandler.

35. Vivasvan Soni, *Mourning Happiness: Narrative and the Politics of Modernity* (Ithaca: Cornell University Press, 2010), 317.

36. McKeon, "Mediation as Primal Word: The Arts, the Sciences, and the Origins of the Aesthetic" in *This is Enlightenment*, ed. Clifford Siskin and William Warner (Chicago: University of Chicago Press, 2010).

37. Jonathan Crary, *24/7: Late Capitalism and the Ends of Sleep* (London: Verso, 2013).

38. Though in this study I use terms drawn from narratology and genre analysis to explore such time that remains, my discussion shares some affinities with Agamben's analysis of messianic time in his *The Time that Remains: A Commentary on the Letter to the Romans* (trans. Patricia Dailey [Stanford: Stanford University Press, 2005]). The enhancing of qualities through representations that highlight composition and sensation—that highlight at once iterability and experience—parallel what Agamben identifies as "operational time"—the very duration of our representing of time. Agamben explains: "Whereas our representation of chronological time, as the time *in which* we are, separates us from ourselves and transforms us into impotent spectators of ourselves—spectators who look at the time that flies without any time left, continually missing themselves—messianic time, an operational time in which we take hold on and achieve our representations of time, is the time *that* we ourselves are, and for this very reason, is the only real time, the only time we have" (68). Agamben discusses the "Romance lyric" (78), and specifically the technique of rhyme, as such a representation that "takes hold and achieves our representations of time"; in the following chapters I demonstrate how numerous narrative technique are geared for this purpose.

39. Frank Kermode, *The Sense of an Ending: Studies in the Theory of Fiction with a New Epilogue* (Oxford: Oxford University Press, 2000), 93–94.

CHAPTER 1

1. Saint Augustine, *Confessions*, trans. Albert C. Outler, rev. trans. Mark Vessey (New York: Barnes and Noble Classics, 2007), book 11. See especially 11.28.38, page 204.

2. Georges Poulet, *Studies in Human Time*, trans. Elliott Coleman (Baltimore: Johns Hopkins University Press, 1956), 21.

3. And Philip Turetzky continues: "Locke clarifies duration as what Descartes called a subjective mode of thought. . . . Apparent time conditions all measurement of duration. Apparent time, then, acts as a boundary condition between the subjective idea of duration and the objective duration of absolute time" (*Time* [London: Routledge, 1998], 78).

4. John Locke, *An Essay Concerning Human Understanding*, ed. Peter H. Nidditch (Oxford: Oxford University Press, 1975). Nidditch relies on the fourth edition of the *Essay*, published in 1700.

5. Jonathan Kramnick points out that Locke inaugurates the modern philosophical approach to consciousness, by which the perception of thinking turns us from living beings into selves or persons. As Kramnick parses such perception of thinking, at stake is what thinking feels like, perhaps indicating the inevitableness of the argument that I make in this section—that Locke's discussion, despite its ostensible focus on how to count time, privileges what time feels like. See *Actions and Objects from Hobbes to Richardson* (Stanford: Stanford University Press, 2010), 85–98. Later in his study Kramnick tracks Locke's revisions to the *Essay* that introduce conscious experience as an important dimension of action. If in the first edition the decision to act is determined mathematically, then "Locke's revisions suggested that one cannot really form a complete picture of actions until one comes to understand the feelings that lead into them: the uneasiness that accompanies desire" (167).

6. Kevis Goodman offers a superb discussion of Locke's and his contemporaries' anxieties about mediation. She identifies the trope of the "microscopic eye," used to underline the difficulties that extraordinary perceptual apparatuses pose for presuming consensus among people's ideas, as well as more profoundly for the very formation of ideas. She argues that without quite acknowledging it, Locke relies on language, and not only on sense perception, to mediate experience and ideas. See *Georgic Modernity and British Romanticism: Poetry and the Mediation of History* (Cambridge: Cambridge University Press, 2004), 42–54.

7. On the invention of the pendulum clock and its cultural impact see Stuart Sherman, *Telling Time: Clocks, Diaries, and English Diurnal Form, 1660–1785* (Chicago: University of Chicago Press, 1996), chapter 1.

8. David Hume, *A Treatise of Human Nature* (Buffalo: Prometheus Books, 1992). Hume's focus on passions and liveliness readily assimilates his work into the tradition of sensibility I have outlined in the Introduction. Annette Baier argues that "Hume enacts for us the turn he wants us to imitate, a turn from a one-sided reliance on intellect and its methods of proceeding to an attempt to use, in our philosophy, *all* the capacities of the human mind: memory, passion and sentiment as well as a chastened intellect" (*A Progress*

*of Sentiments: Reflections on Hume's* Treatise [Cambridge, MA: Harvard University Press, 1991], 1).

9. In his *Abstract* of the *Treatise* Hume declares that "if any thing can intitle the author to so glorious a name as that of an *inventor*, 'tis the use he makes of the principle of the association of ideas, which enters into most of his philosophy" (*An Abstract of a Book Lately Published Entitled* A Treatise of Human Nature [London, 1740], 31). Martin Kallich argues that Hume's discussion deserves such respect for its revolutionary emphasis and complex analysis of associationism, as well as for the way it examines the importance of associationist psychology for aesthetic theory. See *The Association of Ideas and Critical Theory in Eighteenth-Century England: A History of a Psychological Method in English Criticism* (The Hague: Mouton, 1970).

10. For Hume's distinction between substance and mode—two different kinds of complex ideas—see section vi, Book I of the *Treatise.*

11. Gilles Deleuze, *Empiricism and Subjectivity: An Essay on Hume's Theory of Human Nature*, trans. Constantin V. Boundas (New York: Columbia University Press, 1991), 91. Deleuze argues that in Hume we find all the ingredients of a temporal phenomenology. Insofar as Hume defines the mind through its indivisible impressions and its associative operations, Deleuze contends that Hume makes thinking and identity synonymous with time itself; "time was *the structure* of the mind, now the subject is presented as *the synthesis* of time" (94).

12. Baier also underlines the extent to which Hume thoroughly implicates durations and feeling; "The succession of our emotions and passions can be seen not just as feeding into that succession of changing perceptions that clock subjective time for us," she argues, "but as themselves enough to constitute that clock" (51). Baier privileges Hume's discussions of the temporal inflections of our passions in Book II over his more conceptual discussion of time in Book I and argues that his account of moments makes them seem like allegories of persons; "With a bit of exaggeration we could say that the idea of the succession of momentary 'points,' impersonally considered, is derivative from the perceived succession of persons, whose existence is more durable and of whom our perceptions are liveliest" (49).

13. Sean Silver, *The Mind Is a Collection: Case Studies in Eighteenth-Century Thought* (Philadelphia: University of Pennsylvania Press, 2015), 148.

14. Joseph Addison, *The Spectator*, ed. Donald F. Bond, 5 vols. (Oxford: Clarendon Press, 1965).

15. Patricia Meyer Spacks, *Boredom: The Literary History of a State of Mind* (Chicago: University of Chicago Press, 1995), 19.

16. J. H. Plumb, "The Commercialization of Leisure in Eighteenth-Century England," in Neil McKendrick, John Brewer, and Plumb, *The Birth of a Consumer Society: The Commercialization of Eighteenth-Century England* (Bloomington: Indiana University Press, 1982), 265–85; Franco Moretti, "Serious Century," in *The Novel: History, Geography, and Culture*, ed. Moretti, 2 vols. (Princeton: Princeton University Press, 2006), 1: 364–400;

Deidre Shauna Lynch, *Loving Literature: A Cultural History* (Chicago: University of Chicago Press, 2015), 150.

17. Sherman, *Telling Time*, 115.

18. Michael Ketcham, *Transparent Designs: Reading, Performance, and Form in* The Spectator *Papers* (Athens: University of Georgia Press, 1985), 82. Also see Sherman, 146–48.

19. Srinivas Aravamudan, *Enlightenment Orientalism: Resisting the Rise of the Novel* (Chicago: University of Chicago Press, 2012), 20.

20. Elizabeth Deeds Ermarth analyzes a shift in representational conventions from ones informed by a Christian worldview, in which atemporal eternity functions as an organizing principle, to ones informed by sequence and homogeneity underwriting the logic of rational historicism. She finds in early novels a nascent commitment to chronology but argues that by and large they rely on an emblematic form in which plot patterns "yield their meanings typologically, not chronologically" (*Realism and Consensus: Time, Space, and Narrative* [Princeton: Princeton University Press, 1983], 110).

21. Ian Duncan points out the capaciousness of the term "romance" in the eighteenth century: "It is the stuff of the old native minstrelsy, an oriental phantasmagoria, the idle literature of the boudoir, the prophetic and allegorical sentences of protestant enthusiasm and tales of the nursery fireside" (*Modern Romance and the Transformations of the Novel: The Gothic, Scott, Dickens* [Cambridge: Cambridge University Press, 1992], 10).

22. Denis Diderot, *Diderot: Selected Writings on Art and Literature*, ed. and trans. Geoffrey Bremner (London: Penguin, 1994).

23. Jessica Riskin, *Science in the Age of Sensibility: The Sentimental Empiricists of the French Enlightenment* (Chicago: University of Chicago Press, 2002), 1. Riskin quotes here from Fouquet, "Sensibilité, Sentiment" in Diderot and d'Alembert's *Encyclopédie,* and her remarks are about both Diderot and Richardson.

24. Diderot's aesthetics anticipate Viktor Shklovsky's most precisely; he argues that Richardson's lengthy compositions resensitize us to durations that quantitative approaches have habitualized us to in the same way that Shklovsky argues that art resensitizes us to the stoniness of a stone. Shklovsky puts it programmatically, in a way that resonate with my accounts of Locke's and Hume's philosophies as sliding from knowing time to feeling it: "The purpose of art is to impart the sensation of things as they are perceived and not as they are known" ("Art as Technique," rpt. *Modern Criticism and Theory: A Reader*, ed. David Lodge [London: Longman, 1988], 20).

25. In Michael Fried's account of absorption in the age of Diderot, discursive and visual pleasures come into dialogue in ways that cannily resonate with Addison's discussion of methodical composition. Fried argues that at stake is careful attention to composition that solicits a pause from the viewer to contemplate an arresting scene that itself seems to represent a pause in action. When such scenes are well enough crafted, Fried explains, they at once promote an instantaneous response and "set that idea in motion, in dramatic action" (*Absorption and Theatricality: Painting and Beholder in the Age of Diderot* [Chicago:

University of Chicago Press, 1980], 85). Indeed, Fried finds in eighteenth-century French paintings and writings a temporalized defense of absorption that sounds similar to Addison's defense of leisure reading: "Images such as these are not of time wasted but of time *filled* (as a glass may be filled not just to the level of the rim but slightly above)" (51), he writes of Chardin's paintings.

Roger Chartier argues that Diderot's "In Praise of Richardson" instantiates a revolution in reading. Diderot was both an extensive reader—taking advantage of the full range of print publication newly available, while also reading with detachment—and an intensive reader—one who returns again and again to the same beloved text. And as such, Chartier maintains, Diderot selects the best popular leisure print materials and ascribes to them the kind of sentimental-intellectual absorption that he had ascribed to the best paintings. See his *Inscription and Erasure: Literature and Written Culture from the Eleventh to the Eighteenth Century*, trans. Arthur Goldhammer (Philadelphia: University of Pennsylvania Press, 2007), 105–25. For the logic of absorption see especially 105–6 and 116–17.

26. Niklas Luhmann articulates most fully the constitutively social and sensible dimensions of aesthetics in his *Art as a Social System* (trans. Eva M. Knodt [Stanford: Stanford University Press, 2000]). Luhmann defines art as an activity and experience whose function is to dynamically negotiate perception and communication (which he also describes as first order and second order observations, and matter and form). Luhmann valuably indicates how such an understanding of aesthetics was prepared for by eighteenth-century philosophy and criticism, even as, in his view, these early discussions fall short of fully realizing the import of their insights, by and large subordinating communication to perception.

27. Michael McKeon, "Mediation as Primal Word" in *This Is Enlightenment*, ed. Clifford Siskin and William Warner (Chicago: University of Chicago Press, 2010), 409.

## CHAPTER 2

1. Joseph Addison, *The Spectator*, ed. Donald F. Bond, 5 vols. (Oxford: Clarendon Press, 1965).

2. Drawing on G. A. Starr and J. Paul Hunter, Elizabeth Deeds Ermarth explains that Christian temporal form highlights emblematic significances—the moment as it resonates with eternity—rather than historical significances—the moment as it is positioned within a sequence of linear unfolding. See Ermarth's *Realism and Consensus in the English Novel: Time, Space, and Narrative* (Princeton: Princeton University Press, 1983).

3. Lewis White Beck, "World Enough, and Time" in *Probability, Time, and Space in Eighteenth-Century Literature*, ed. Paula R. Backscheider (New York: AMS Press, 1979), 114.

4. For Hutcheson's utilitarianism, see Stephen Darwall, *The British Moralists and the Internal "Ought": 1640–1740* (Cambridge: Cambridge University Press, 1995), 216. T. D. Campbell also emphasizes Hutcheson's utilitarianism, as well as his position within the Enlightenment tradition; see "Francis Hutcheson: 'Father' of the Scottish Enlightenment,"

in *The Origins and Nature of the Scottish Enlightenment*, ed. R. H. Campbell and Andrew S. Skinner (Edinburgh: John Donald, 1982), 167–85. From these studies I learned much about the complexities of Hutcheson's oeuvre and about his position within the tradition of the Scottish Enlightenment. However, these do not attend to the durational stakes of Hutcheson's arguments.

5. Denis Diderot, "In Praise of Richardson," in *Diderot: Selected Writings on Art and Literature*, ed. and trans. Geoffrey Bremner (London: Penguin, 1994), 86.

6. Ian Watt, *The Rise of the Novel: Studies in Defoe, Richardson, and Fielding* (Berkeley: University of California Press, 1957); Stuart Sherman, *Telling Time: Clocks, Diaries, and English Diurnal Form, 1660–1785* (Chicago: University of Chicago Press, 1996).

7. Letter to Lady Bradshaigh, 9 October 1756; in *Selected Letters of Samuel Richardson*, ed. with intro. John Carroll (Oxford: Oxford University Press, 1964), pg. 329.

8. Mark Kinkead-Weekes, *Samuel Richardson: Dramatic Novelist* (Ithaca: Cornell University Press, 1973), 401. Kinkead-Weekes makes the case more generally for the range of techniques through which Richardson's epistolary form achieves psychological authenticity. Tom Keymer, contrary to Kinkead-Weekes, argues that Richardson's epistolary form conveys the complexity of the art of writing, rather than represents truth—especially not truth about the interiority of character. Keymer, *Richardson's* Clarissa *and the Eighteenth-Century Reader* (Cambridge: Cambridge University Press, 1992).

9. Leah Price, *The Anthology and the Rise of the Novel: From Richardson to George Eliot* (Cambridge: Cambridge University Press, 2000), 16.

10. Letter to Lady Bradshaigh, 14 February 1754 in *Selected Letters*, 289. Carroll prints additions and deletions Richardson made to copies he kept of his letters; phrases omitted by Richardson in these revisions are place within angle brackets, as <that> in the quotation above.

11. Vivasvan Soni, *Mourning Happiness: Narrative and the Politics of Modernity* (Ithaca: Cornell University Press, 2010), 315–16.

12. Soni, *Mourning Happiness*, 304. Soni understands Smith's and Richardson's work as transitional. They hesitate, he argues, between a Classical tragic ethic of pity—concern for the happiness of others through attention to their narrative situation—and a modern ethic of identification—a narcissistic imagining of ourselves feeling like another, through attention to tableaus of present conditions. But in the final analysis, Soni contends, Smith and Richardson privilege identification.

13. Hina Nazar, *Enlightened Sentiments: Judgment and Autonomy in the Age of Sensibility* (New York: Fordham University Press, 2012).

14. James Chandler, *An Archaeology of Sympathy: The Sentimental Mode in Literature and Cinema* (Chicago: University of Chicago Press, 2013), Part 2 "The Making of Literary Sentimentalism," and especially 217–22.

15. For the theatrical model and the transformation of Hume's thought, see John Mullan, *Sentiment and Sociability: The Language of Feeling in the Eighteenth Century* (Oxford: Oxford University Press, 1988), 36–44. Mullan argues that Hume's account of sympathy in the *Treatise of Human Nature* (1739) differs substantially from his accounts in the

*Enquiries*; whereas the latter highlight distance and position, the former conceptualizes fellow-feeling through common vibrations, thus underlining not spectatorship but tactile and aural experience. For extensive studies of theatrical sympathy in the eighteenth century, see David Marshall, *The Figure of Theater: Shaftesbury, Defoe, Adam Smith, and George Eliot* (New York: Columbia University Press, 1986) and *The Surprising Effects of Sympathy: Marivaux, Diderot, Rousseau, and Mary Shelley* (Chicago: University of Chicago Press, 1988). I discuss the differences between theatrical and tactile models of sympathy in the next chapter.

16. See Nazar 49–50 and Chandler 236–37.

17. Francis Hutcheson, *An Essay on the Nature and Conduct of the Passions and Affections, with Illustrations on the Moral Sense*, ed. and intro. Aaron Garrett (Indianapolis: Liberty Fund, 2002).

18. Hutcheson, *An Inquiry into the Original of Our Ideas of Beauty and Virtue in Two Treatises* ed. and intro. Wolfgang Leidhold, rev. ed. (Indianapolis: Liberty Fund, 2008). Leidhold uses Hutcheson's revised second edition from 1726 and, thus, so do I; Hutcheson published the first edition a year earlier.

19. *Inquiry*, throughout, but see especially 104 and 88–100, *Illustrations* 133–36.

20. In a reading that emphasizes passivity, Scott Paul Gordon makes the case for the moral sense's immediacy, as it subtends reasoned deliberation that, he believes, necessarily promotes self-interested calculations. He underlines Hutcheson's polemic against prudential theories such as Hobbes's and Mandeville's, explaining that Hutcheson privileges passive motivation, which in the very speed and passion of its manifestation precludes the possibility of self-interested calculations. See *The Power of the Passive Self in English Literature, 1640–1770* (Cambridge: Cambridge University Press, 2002), 119–52, especially 127 and 133.

Daniel Carey describes Hutcheson's moral sense as a "faculty or capacity, which combined the moment of moral perception with that of moral judgement," in *Locke, Shaftesbury, and Hutcheson: Contesting Diversity in the Enlightenment and Beyond* (Cambridge: Cambridge University Press, 2006), 151. Carey seems simply to mean that Hutcheson designs his moral sense to ensure an instinctive approval of benevolent action prior to any computations of personal advantage. But in characterizing perception and judgment as two different moments, rather than just two different activities, he signals both the high incidence of the word "moment" in Hutcheson's writings, and the moral sense's compression into a single instant of what we may take as quite long and separate time spans. I use "signals," rather than "flags," because Carey doesn't elaborate on this problematic durational aspect. Nonetheless, his formulation seems to me uncannily perceptive, pointing out the need to consider not only cultural but also temporal diversity in Hutcheson's work.

21. Carey helpfully tracks Hutcheson's various articulations of this universal capacity, explaining where and how he defines the moral sense as a faculty—already given fully formed—and where and how he articulates it as a predisposition—requiring experience in order to develop (162–71).

22. See Tony C. Brown, *The Primitive, the Aesthetic, and the Savage: An Enlightenment*

*Problematic* (Minneapolis: University of Minnesota Press, 2012). Darwall addresses the question of how a capacity engages the phenomenal world by explaining that the kind of "internalism" we find in Hutcheson assumes that if a moral proposition is true—or an action good—then a motive arises in us on its behalf. It is internal to the agent yet it also validates the proposition. *The British Moralists and the Internal "Ought,"* 11.

23. This leads Darwall to argue that the moral sense can only be an observational stance: "According to Hutcheson, it is only the moral sense of *observers* that is pertinent to virtue, enabling them to appreciate the moral goodness of first-order moral motives (all forms of benevolence, for Hutcheson). Moral sense's only relevance from an *agent's* perspective, in deliberation, is to exhibit a self-interested concern in being virtuous, namely, to achieve the pleasures of reflecting on one's own virtue. And this motive cannot enhance the agent's virtue" (209). Furthermore, Darwall argues that through changes Hutcheson makes to subsequent editions of the *Inquiry*, we can see that what ultimately motivates action is not the moral sense but, rather, a sense of benevolence. To me it seems that Hutcheson's various discussions of the moral sense and the sense of benevolence are sufficiently similar to identify the two and thus to understand the moral sense as both observational and motivational.

24. Jeremy Bentham, *The Principles of Morals and Legislation* (1781; rpt. Amherst, NY: Prometheus Books, 1988), 2.

25. On cultural diversity, see Carey, who examines how eighteenth-century philosophers respond to the problem as it is occasioned by Locke's argument against innateness. On Carey's account, Shaftesbury reacts against Locke's rejection of innateness and against his imagining of human motivation as founded on appetites and self-interest; Hutcheson tries to synthesize Locke's and Shaftesbury's opposing views. On doctrinal diversity, see Isabel Rivers, *Reason, Grace, and Sentiment: A Study of the Language of Religion and Ethics in England 1660–1780*, 2 vols. (Cambridge: Cambridge University Press, 2000).

26. In a stratified society characterized by inequality, the utilitarian formula attends to status as an essential variable. But Hutcheson emphasizes that while access to ability varies, access to benevolence is equal; as he continues to explain his formula for the moment of good in the *Inquiry*, he highlights the universality of the moral sense: "No external Circumstances of Fortune, no involuntary Disadvantages, can exclude any Mortal from the most heroick Virtue" (134).

27. But not in judgments of legal culpability—as we've seen earlier and as Hutcheson occasionally makes clear in comments throughout the *Treatise* and the *Illustrations*.

28. See Watt, *The Rise of the Novel*, especially chapters 6–8.

29. Jonathan Kramnick, *Actions and Objects from Hobbes to Richardson* (Stanford: Stanford University Press, 2010) and Sandra Macpherson, *Harm's Way: Tragic Responsibility and the Novel Form* (Baltimore: Johns Hopkins University Press, 2010).

30. Samuel Richardson, *Clarissa, or, The History of a Young Lady* (1747–1748), ed. and intro. Angus Ross (Harmondsworth, UK: Penguin, 1986). Ross reprints the text of Richardson's first edition. Although some scholars prefer relying on Richardson's third edition, as reprinted in AMS Press's *The Clarissa Project*, I agree with Ross's prioritizing of

accessibility and enjoyment, which are better figured in the first edition; I therefore work with the Penguin version.

31. The lengthy exchanges between Anna and Clarissa in the early parts of the novel about the best course of action given rapidly changing circumstances, and, later in the novel, Clarissa's presentation to Belford and Anna of her reasons for declining to marry or sue Lovelace and for preparing for her death vindicate Clarissa, despite the strangeness of her choices and their catastrophic outcomes. See my "Reasonableness and Domestic Fiction" (*ELH* 73.4 [2006]: 85–30) and Nazar's chapter 3 of *Enlightened Sentiments*. Also see Tom Keymer's *Richardson's Clarissa*, where he identifies a casuistic model in Richardson's writing—a malleable relation between case and rule, which makes judgment a deliberative project.

32. Samuel Richardson, *Pamela; or, Virtue Rewarded* (1740), ed. Thomas Keymer and Alice Wakely, intro. Keymer (Oxford: Oxford University Press, 2001).

33. The code of love as passion, Luhmann points out, makes temporality a key dimension, specifically by casting deferral as the relationship's condition of possibility and the moment of actualization as a pivotal turning point towards dissolution (*Love as Passion: The Codification of Intimacy*, trans. Jeremy Gaines and Doris L. Jones [1986; rpt. Stanford: Stanford University Press, 1998], 70–75). This puts women, he notes, at a disadvantage and makes rife conditions for a conceptualization of virtue as calmness and stability (74). Luhmann takes the English privileging of the moral logic over the passionate logic to entail the failed integration of sexuality with love—the rise of the companionate marriage inevitably eventuates in Victorian prudishness (114), he argues—as well as a sharpening of the divide between domestic intimacy and public impersonality (100). This conclusion seems hardly disputable, and yet it may be precisely because English sensibility does not rely on sexual consummation as the crux of its code for intense emotional attachment that in it the burden of intimacy's uniqueness falls almost entirely on an alternate temporality. Moreover, because it distances love as passion from sexuality, English sensibility can extend its code of love as a passionate duration to sociability and sympathy more generally. In the French case, the code of love as passion had a far-reaching historical impact insofar as its logic came to dominate modern conceptions of sexual intimacy, ultimately also coming to define ideals of marriage—"libertinage as a case of evolutionary good fortune, if you will" (6), as Luhmann puts it. The English case may have also had its version of "libertinage as a case of evolutionary good fortune," but here it involved not the reintegration of sex into marriage, but the importation of qualitative duration out to sympathy. I will say more about the duration of sympathy in the next chapter.

34. Pamela's and B's wedding negotiations instantiate especially well the kind of personalization of calendar that Stuart Sherman makes the case for—the "plotting of narratives and subjectivities within and against the grid of successive, dated days" (*Telling Time*, 247). Pamela asks B to marry on a Thursday because her family history seems to have had many happy Thursdays: "I have, Sir, proceeded I, a great Desire, that whenever the Day is, it may be of a *Thursday*: Of a *Thursday* my dear Father and Mother were marry'd, and tho' poor, they are a very happy Pair.—Of a *Thursday* your poor *Pamela* was born: Of a *Thursday* my dear good Lady took me from my Parents into her Protection: Of a *Thursday*, Sir,

you caus'd me to be carry'd away to this Place, to which I now, by God's Goodness and your Favour, owe so amazingly all my present Prospects; and of a *Thursday* it was, you nam'd to me that Fourteen Days from that, you would confirm my Happiness" (326). In response, B comments, "This, *Pamela*, is a little superstitious" (326); he offers to make Monday a new benchmark in Pamela's calendar and adds, "I hope, we shall make *Tuesday, Wednesday, Friday, Saturday* and *Sunday*, as happy Days, as *Monday* and *Thursday* (327). Yet when negotiations over Monday commence, it becomes clear that B has as much attachment to his personal temporal experience as does Pamela: "What! A Week still? said he. Sir, answer'd I, if you please; for *that* will be, as you injoin'd, within the second Seven Days. Why, Girl, said he, 'twill be Seven Months till next *Monday*" (327). In his raillery of Pamela's attachment to a particular day of the week, B suggests the irrelevance of private experience to the public calendar. Yet, in specifying other weekdays as a new conjugal project and betraying that for him seven days can be as seven months, he indicates how the calendar is a backdrop for his interpersonal negotiations, as well as for his personal sense of time.

35. Lovelace's libertinism is, thus, a far cry from Luhmann's libertinism of love as passion. Luhmann's account suggests that the growth of time—actively making time grow—is precisely the business of love as passion in its consisting of narrative delays to forestall the inevitable dissolution of actualization.

36. William B. Warner, "Proposal and Habitation: The Temporality and Authority of Interpretation in and about a Scene of Richardson's *Clarissa*" *Boundary 2* 7.2 (1979), 170.

37. Thomas M. Kavanagh finds in eighteenth-century French theorizations of aesthetic experience an idealization of unfettered moments, which resonates with Lovelace's libertinism as I have described it: "a primary and iconoclastic preference not to the uninterrupted confluences of present with past and future but to the singular power of the moment as a temporality standing outside of and defining itself against those assumed coherencies" (*Esthetics of the Moment: Literature and Art in the French Enlightenment* [Philadelphia: University of Pennsylvania Press, 1996], 7).

38. See Michel Foucault, *Discipline and Punish: The Birth of the Prison*, trans. Alan Sheridan (Harmondsworth, UK: Penguin, 1977).

39. Sherman, *Telling Time*, 19, 21.

40. Jesse Molesworth, *Chance and the Eighteenth-Century Novel: Realism, Probability, Magic* (Cambridge: Cambridge University Press, 2010), 7. Molesworth argues that the popularity of teleological plots in eighteenth-century England indicates the increasing commitment to fictionalizing modes of thinking, rather than any attempt to achieve a more accurate verisimilitude in representation (realism).

41. In *Narrative Discourse Revisited*, trans. Jane E. Lewin (Ithaca: Cornell University Press, 1988), Genette expands his earlier specification of pauses in *Narrative Discourse* to distinguish between a descriptive pause and a reflexive pause. He explains: "In those cases [of reflexive pause] it cannot be said, as for the descriptive pause, that the narrative slows down by bringing the time of its story to a standstill so as to cast a glance over its diegetic space. Rather, it interrupts itself to give up its place to another type of discourse" (37). On descriptive pauses as constituting a key dimension of the expansive narrative middles of

Victorian novels, see Amy King, "Dilatory Description and the Pleasures of Accumulation: Toward a History of Novelistic Length" in *Narrative Middles: Navigating the Nineteenth-Century British Novel*, ed. Caroline Levine and Mario Ortiz-Robles (Columbus: Ohio State University Press, 2011), 161–94.

42. On digression as a technique of plot in *Tristram Shandy*, see Keymer, *Sterne, the Moderns, and the Novel* (Oxford: Oxford University Press, 2002). I will discuss *Tristram Shandy*'s digressive mode in the next chapter.

43. Caroline Levine, "An Anatomy of Suspense: The Pleasurable, Critical, Ethical, Erotic Middle of *The Woman in White*" in *Narrative Middles*, 197.

44. For a discussion of adventure and error as principles of romance, see the Introduction.

45. Molesworth argues that novels' plots highlight plausibility or possibility, rather than probability.

46. Deidre Shauna Lynch, *The Economy of Character: Novels, Market Culture, and the Business of Inner Meaning* (Chicago: University of Chicago Press, 1998), 218. Though at stake is deciphering meaning in Lynch's account here, which she further develops in *Loving Literature* with emphasis on the iterative, such literary experience of reading temporally and iteratively evokes feeling: "while they meditate on what it means to repeat an experience or a story with renewed feeling, her [Austen's] novels vividly annotate the practices of romantic readers" (*Economy*, 220).

47. See Soni, "Committing Freedom: The Cultivation of Judgment in Rousseau's *Emile* and Austen's *Pride and Prejudice*," *The Eighteenth Century* 51.3 (2010): 363–87. Soni focuses on the judgment of character; he demonstrates that one has to come up with a revised sense of character—hypothesis about the logic of a person—in order to read the facts of action in such a way that would confirm this new assessment. Thus determining which logic of character and which details of action ought to be privileged depends on narrative time. Therefore, what Austen produces—and what we can only have, according to Soni, if judgment is free—is a phenomenology of judgment, rather than a theory of it.

48. Jane Austen, *Sense and Sensibility*, ed. with intro. Ros Ballaster (New York: Penguin, 1995).

## CHAPTER 3

1. *Letters of Laurence Sterne*, ed. Lewis Perry Curtis (Oxford: Clarendon Press, 1935). These are the very first lines of the letter, which begins in medias res. As Curtis explains in his notes to this edition, this letter—number 76—has no date on it; thus the June 20, 1761, date is a likely attribution, based on the timeline of Sterne's and Mrs. Vesey's acquaintance (138–39).

2. Arthur H. Cash maintains that the letter, with its audacious puns, was probably never sent to its addressee: "If Sterne had been foolish enough actually to send his letter, it would have instantly put an end to any budding romance or friendship. No, the probabil-

ities are that Elizabeth Vesey meant little or nothing to Sterne except as a figure in some fantasy which he transmogrified into a piece of comic writing" (*Laurence Sterne: The Later Years* [London: Methuen, 1986], 27).

3. Adam Smith, *The Theory of Moral Sentiments* (1759); ed. D. D. Raphael and A. L. Macfie (Oxford: Oxford University Press, 1976; rpt. Indianapolis: Liberty Press, 1982).

4. G. J. Barker-Benfield, *The Culture of Sensibility: Sex and Society in Eighteenth-Century Britain* (Chicago: University of Chicago Press, 1992); the quotation is from Cheyne's *The English Malady* (1733; rpt. Delmar, N.Y.: Scholars Facsimiles and Reprints, 1976), 3–4. R. F. Brissenden in *Virtue in Distress: Studies in the Novel of Sentiment from Richardson to Sade* (London: Macmillan, 1974) and Ann Jessie Van Sant in *Eighteenth-Century Sensibility and the Novel: The Senses in Social Context* (Cambridge: Cambridge University Press, 1993), also emphasize sensibility's vibratory model of perceptual processes.

5. Glenn Gould's interview with Tim Page is reproduced as disc 3 of *A State of Wonder: The Complete Goldberg Variations,* the 2002 Sony disc set which also includes the 1955 and 1981 recordings.

6. *Roger North's The Musicall Grammarian, 1728*, ed. with intro. and notes Mary Chan and Jamie C. Kassler (Cambridge: Cambridge University Press, 2006).

7. See Kassler, *Inner Music: Hobbes, Hooke and North on Internal Character* (London: Athlone, 1995), esp. 1–4, 130–37, 168–74. Kassler's work on early English musicology generally and on Roger North particularly is invaluable. It includes annotated editions of North's work (in addition to *The Musicall Grammarian*, see *Roger North's Cursory Notes of Musicke (c. 1698–c. 1703); A Physical, Psychological, and Critical Theory*, ed. Chan and Kassler [Kensington: Unisearch, 1986]) and a comprehensive bibliography of all pamphlets, essays, and treatises on music published in England during the period, *The Science of Music in Britain, 1714–1830: A Catalogue of Writings, Lectures and Inventions* 2 vols. (New York: Garland, 1979).

8. Quoted in Kassler, *Inner Music*, 138, 171.

9. Letter to *Quarterly Musical Magazine and Review* in 1827; quoted in Kassler, *The Science of Music in Britain, 1714–1830*, 1:516. According to various entries in Kassler's catalogue, during the first decade of the nineteenth century there was a lively debate about which mechanical device is better, stopwatch or pendulum, and about when and where these technologies were first introduced.

10. According to Wilbur Samuel Howell, by the close of the century, the elocutionists were considered authorities on pronunciation, their influence arising from the need for a standardized medium for a consolidating nation aspiring to govern a growing empire. See *Eighteenth-Century British Logic and Rhetoric* (Princeton: Princeton University Press, 1971), 156–59. Howell argues that the elocutionists' work was by and large a didactic enterprise aimed at preparing men of provincial origin to succeed in London. For such disciplinary understanding of the elocutionist enterprise, also see Michael Shortland's "Moving Speeches: Language and Elocution in Eighteenth-Century Britain," *History of European Ideas* 8:6 (1987): 67–88.

11. Though it is by no means limited to their discussions of rhythm. Julia Carlson

beautifully tracks how the elocutionist project of mapping and notating speech makes available subtle variations in melodiousness, rather than flattening and regularizing it; see her *Romantic Marks and Measures: Wordsworth's Poetry in Fields of Print* (Philadelphia: University of Pennsylvania Press, 2016), especially 129–44. The elocutionists were key to what Murray Cohen has identified as the mid-eighteenth-century renewed interest in the sensory components of language. Cohen explains that the work of linguists from the second half of the eighteenth century approaches language as "a collection of native habits that could be compiled or used as evidence of forces like history, geography, and the progress of human societies, or could be studied rhetorically as effective and affective speech" (*Sensible Words: Linguistic Practice in England, 1640–1785* [Baltimore: Johns Hopkins University Press, 1977], xxv).

12. See Thomas Sheridan's *A Course of Lectures on Elocution* (London, 1762) and John Herries' *The Elements of Speech* (London, 1773). For more about Sheridan's approach to rhythm, see my "The Sense of Rhythm: Nationalism, Sympathy, and the English Elocutionists," *The Eighteenth Century: Theory and Interpretation* 52.2 (2011): 173–92. Sheridan was probably the most influential elocutionist among his contemporaries and the most extensively studied by recent scholars.

13. Joshua Steele, *An Essay Towards Establishing the Melody and Measure of Speech* (1775; rpt. Menston, England: Scolar Press, 1969). The *Essay* was written as a polemic against James Burnett, Lord Monboddo's *Of the Origin and Progress of Language*, and includes Steele's responses to Monboddo's rebuttals made in a private correspondence between the two. According to David Abercrombie, Steele managed to persuade Monboddo of all his points. See *Studies in Phonetics and Linguistics* (Oxford: Oxford University Press, 1965), 33–44. Steele's *Essay* was well received at publication and was reissued with additions in 1779 under the title *Prosodia Rationalis*. For accounts of Steele's reception, see Paul K. Alkon, "Joshua Steele and the Melody of Speech," *Language and Speech* 2:3 (1959): 154–74 and Kassler, "Representing Speech through Musical Notation," *Journal of Musicological Research* 24.3–4 (2005): 227–39.

14. Abercrombie, 39.

15. Many scholars remark on the nationalist motivations of the elocutionary project. Nicholas Hudson, for example, argues that Sheridan cultivates voice and oral delivery as essential to national health, antidotes to the increasing influence of print media (*Writing and European Thought, 1600–1830* [Cambridge: Cambridge University Press, 1994], 103–10 and esp. 107–9). And Jay Fliegelman explains that the elocutionists' revision of rhetoric informed conceptions of political authority in the new American republic, and mentions in passing Hume's discussion of sympathy as yet another English source for the "affective consent" underwriting the new republic's legitimate sovereignty. See *Declaring Independence: Jefferson, Natural Language, and the Culture of Performance* (Stanford: Stanford University Press, 1993), 35–36, 40, and more generally 35–62, for how "affective consent" arises from both a Humean conception of sympathy and the elocutionists' transformation of oratory.

16. Herries, *The Elements of Speech*.

17. Sheridan, *A Course of Lectures on Elocution*, 56.

18. Sheridan in the quotation above uses "ear" literally and "tongue" metaphorically. But Herries argues that "tongue" as a metaphor for language also has its literal underpinnings: he traces this figuration to the physical functioning of the organ while pronouncing words (17).

19. Earlier studies of the elocutionists have taken the body as a site of resistance to formalization. Thus Peter De Bolla argues that for many elocutionists the body finally reveals "the unlegislatable"; "The body may say something else," the elocutionists worried, according to De Bolla, "a fear that the language of the body may be excessive, that it may produce a legible, but far from wanted surplus." See *The Discourse of the Sublime: Readings in History, Aesthetics, and the Subject* (Oxford: Basil Blackwell, 1989), 159, 161. While such clashes may certainly happen, I have been showing the extent to which the elocutionists' accounts of rhythm take pains to conceive of a form adequate to bodies' tendencies, rather than indifferent to these. In her careful and nuanced tracking of elocutionists' discussions of tonal sliding and emphasis, Carlson offers an even more extensive case than I do for the elocutionists' sophisticated approach to regularization and abstraction and for their commitment to democratization. See *Romantic Marks and Measures*, interchapter, and chapters 4 and 5.

The elocutionists' embracing of modern projects of nation and standardization distinguishes them from Jean-Jacques Rousseau's understanding of language's musicality. While Steele and Sheridan identify rhythm as a condition of language as such, and consider modernity no less capable of realizing its sympathetic potentials, Rousseau in his *Essay on the Origin of Languages* (1781) considers musicality and fellow-feeling as qualities of long-past innocent cultures and languages. And while both the elocutionists and Rousseau associate the musicality of language with presence, the former imagine ways in which such presence can be conveyed and elicited in writing, whereas Rousseau suspects writing of irremediably eradicating the possibility of full presence in language.

20. For a much fuller explanation of Steele's musical notation, see Kassler, "Representing Speech," esp. 233–35 for notation of rhythm.

21. Andrew Elfenbein, *Romanticism and the Rise of English* (Stanford: Stanford University Press, 2009), 113, and see more generally 111–17. For an extensive analysis of the elocutionists' typographical innovations, see Carlson.

22. Such distant representation of others' feelings, according to Marshall, opens the possibility of misunderstanding and alienation; sympathy posited on representation may, in the final analysis, produce precisely its opposite—encouraging individuation and undercutting community—a contradiction anxiously haunting, according to Marshall, much eighteenth-century writing on fellow-feeling. As Marshall puts it, "In reading or beholding the characters of others, one risks not only being misled but also being placed in the position of distance, difference, and isolation that sympathy is supposed to deny" (*The Surprising Effects of Sympathy: Marivaux, Diderot, Rousseau, and Mary Shelley* [Chicago: University of Chicago Press, 1988], 181). Marshall similarly concludes his discussion of Smith's *The Theory of Moral Sentiments*: "The dream of sympathy, the fiction of sympathy, is that an

interplay and interchange of places, positions, persons, sentiments, and points of view could cancel out the theatricality of the most theatrical of situations" (*The Figure of Theater: Shaftesbury, Defoe, Adam Smith, and George Eliot* [New York: Columbia University Press, 1986], 192). I take Marshall's choice of "dream" and "fiction" here to indicate the unreality of sympathy's aspiration to cancel theatricality.

23. William Wordsworth, *Lyrical Ballads*, ed. Michael Mason. 2nd ed. (Harlow, UK: Pearson Longman, 2007).

24. But there are some notable exceptions to this tendency to focus on meter. Elfenbein demonstrates the extent to which beat underwrites the effectiveness of some romantic poetry, which he calls "percussive romanticism," and Carlson tracks the ways in which emphasis and punctuation, no less than meter, inform Wordsworth's conception of his poetic medium.

25. While many scholars have taken Locke as a major influence on Sterne, I follow the lead of John Traugott, who in *Tristram Shandy's World: Sterne's Philosophical Rhetoric* (Berkeley: University of California Press, 1954) makes a persuasive case against understanding Sterne's paraphrases of Locke as straightforward adaptations of an authoritative source. Instead, he points to Hume and the associationists as more relevant influences. Jonathan Lamb in *Sterne's Fiction and the Double Principle* (Cambridge: Cambridge University Press, 1989) agrees with and expands on Traugott.

26. John Locke, *An Essay Concerning Human Understanding*, ed. Peter H. Nidditch (Oxford: Oxford University Press, 1975).

27. Laurence Sterne, *The Life and Opinions of Tristram Shandy, Gentleman* (1759–67), ed. Melvyn New and Joan New (Gainesville: University Press of Florida, 1978–84), vols. 1–3 of *The Florida Edition of the Works of Laurence Sterne.*

28. See Lamb, *Sterne's Fiction and the Double Principle.* Lamb focuses on the skepticism generated by a doubleness of literal and figurative meaning: Sterne is "fascinated by the confusion of the intellectual and the material worlds that comes about when words straddle the border between figurative and literal meanings. Like the transformations it accompanies, this wordplay enacts a repudiation of the thing itself" (30).

29. More generally, Sterne's writings are, as Alexis Tadié puts it, "vocal and auditory productions" (*Sterne's Whimsical Theatres of Language: Orality, Gesture, Literacy* [Burlington: Ashgate, 2003], 20). Tadié considers the complexity of Sterne's engagement with the various material dimensions of language and with the semiotic dimensions of gestures. For studies of music and Sterne, also see William Freedman *Laurence Sterne and the Origins of the Musical Novel* (Athens: University Georgia Press, 1978) and John Leslie "Music's Sentimental Role in *Tristram Shandy*" *Papers on Language and Literature* 41.1 (2005): 55–66.

30. For Stuart Sherman's analysis of the dialectic of diurnal form in travel writing, see *Telling Time: Clock, Diaries, and English Diurnal Form, 1660–1785* (Chicago: University of Chicago Press, 1996), chapter 5. For a discussion of the complex ways in which Sterne challenges conventions of travel writing, see Tadié, 111–29.

31. Carlson considers the elocutionists' reconceptualization of punctuation as a

technique of pacing geared better to convey voice and feeling; punctuation—dashes, exclamation marks, commas, periods—marks pauses and these, the elocutionists argue, are necessary to give sufficient time to let ideas sink in and evoke emotion.

32. See Lynn Festa's evocative discussion in *Sentimental Figures of Empire in Eighteenth-Century Britain and France* (Baltimore: Johns Hopkins University Press, 2006) of the importance of repetition in sentimental fiction. Festa argues that sentimental language thrives on both recurring use and individualized appropriation through narrative, though her account underlines tensions between individual experience and iteration.

33. On Sterne's fascination with Garrick as master of sensibility, see John Mullan's *Sentiment and Sociability: The Language of Feeling in the Eighteenth Century* (Oxford: Oxford University Press, 1988), 175. On Sterne's use of theatrical conventions in his novels, see Tadié, 77–108.

34. Janine Barchas, *Graphic Design, Print Culture, and the Eighteenth-Century Novel* (Cambridge: Cambridge University Press, 2003), 161. Barchas lists the various significations of the dash: "It marks pauses, rhetorical transitions, approximate syntax, and moments of aposiopesis—the intentional refusals to complete an idea, name or phrase so common in eighteenth-century literature" (158). Though for Barchas most important are instances of such aposiopesis where language seems to fail, for my purposes more important is pacing of successful language through pauses and transitions.

35. Mullan, *Sentiment and Sociability*, 160.

36. Lamb, *Sterne's Fiction and the Double Principle*, 15.

37. Lamb, 64, 65.

38. See Lamb, 64–73. Among the many examples Lamb discusses, Trim's "puffs" are my favorite: "The word 'puff' resonates in the same way when Trim sets out to mimic the cannonade of Lille with two Turkish tobacco pipes fitted to a set of miniature brass cannon. 'His first intention, as I said, was no more than giving the enemy a single puff or two;---but the pleasure of the *puffs*, as well as the *puffing*, had insensibly got hold of the corporal, and drawn him on from puff to puff' (TS [*Tristram Shandy*], 6.27.454). In these circumstances there is no such thing as a single puff, for each comprises an impression and an idea: the impression of literal smoking and the idea of figurative gunnery" (73).

39. Thomas Keymer considers the ways by which Sterne makes this point at the level of content; he perceptively notes that if in early volumes Sterne describes narration's losing battle against action in terms of "time pressing," in later ones he describes it in terms of "time wasting" (*Sterne, the Moderns, and the Novel* [Oxford: Oxford University Press, 2002], 143). Everett Zimmerman offers another evocative discussion of Sterne's representation of mortality in his "From Personal Identity the Material Text: Sterne, Mackenzie, and Scott" in his *The Boundaries of Fiction: History and the Eighteenth-Century British Novel* (Ithaca: Cornell University Press, 1996), 179–224.

40. Keymer, *Sterne, the Moderns, and the Novel*, 9.

41. Gérard Genette, *Narrative Discourse: An Essay in Method*, trans. Jane E. Lewin (Ithaca: Cornell University Press, 1980).

42. Genette's narratological terms for these four movements are ellipsis, descriptive pause, scene, and summary. In *ellipsis* a "nonexistent section of narrative corresponds to some duration of story" (93); in *descriptive pause* "some section of narrative discourse corresponds to a nonexistent diegetic duration" (93–94); and between these two extremes we find *scene*: "most often in dialogue, which, as we have already observed, realizes conventionally the equality of time between narrative and story" (94), and *summary*: "a form with variable tempo (whereas the tempo of the other three is fixed, at least in principle), which with great flexibility of pace covers the entire range included between scene and ellipsis" (94).

43. Michael Levenson, "Reading Time," *Novel: A Forum on Fiction* (2009) 42.3: 511–16. For narratology's continued privileging of chronology and order, see Shlomith Rimmon-Kenan, who while explicitly relying on Genette, announces at the outset of her consideration of "Text: Time" that she will focus on "time as the textual arrangement of the event component of the story" (*Narrative Fiction: Contemporary Poetics* [London: Methuen, 1983], 43).

44. Jane Austen, *Sense and Sensibility*, ed. with intro. Ros Ballaster (New York: Penguin, 1995).

45. Genette, *Narrative Discourse Revisited*, trans. Jane E. Lewin (Ithaca: Cornell University Press, 1988).

46. Garrett Stewart, *Novel Violence: A Narratography of Victorian Fiction* (Chicago: University of Chicago Press, 2009), 6; emphasis in the original. Stewart explains that the narratologies of Greimas, Bakhtin, Genette, Ricoeur, Todorov, and even Barthes are less evocative precedents for his purposes and turns, instead, to Roman Jakobson's poetics.

47. In passing, Stewart recognizes that a narratographic analysis of eighteenth-century novels would yield a different account of the relation between prose and the story it tells. "Though Fielding's plot sends Jonathan Wild straight to the gallows, prose never gets inside the inevitable terror of the moment," and "within the abstracted confines of such poetic justice, nothing human feels really violated" (23). My reading of *Tristram Shandy* suggests, however, that while indeed "nothing human feels really violated," this is not because Sterne's prose is abstracted from the stories it tells, but because the "inside" that the prose of narration reveals is not cast as terror, but rather as comic and ordinary human immersions.

## CHAPTER 4

1. Christopher Miller, *The Invention of Evening: Perception and Time in Romantic Poetry* (Cambridge: Cambridge University Press, 2006).

2. Ian Duncan reminds that Radcliffe was "the first novelist to command huge fees from her publisher and the name-recognition of a middle-class reading public" and she "pioneered the cultural position of the author as famous producer of a certain

commodity that would be inherited by Scott and bequeathed as canonical to the nineteenth century" (*Modern Romance and the Transformations of the Novel: The Gothic, Scott, Dickens* [Cambridge: Cambridge University Press, 1992], 35). In considering Radcliffe's contribution to the refiguration of romance, Duncan focuses on her revision of paternal archetypes and anxieties.

3. Mikhail Bakhtin, "Forms of Time and of the Chronotope in the Novel," *The Dialogic Imagination: Four Essays*, ed. Michael Holquist, trans. Caryl Emerson and Michael Holquist (Austin: University of Texas Press, 1981), 152.

4. John Locke, *An Essay Concerning Human Understanding*, ed. Peter H. Nidditch (Oxford: Oxford University Press, 1975), 185.

5. Edmund Burke, *A Philosophical Enquiry into the Origin of Our Ideas of the Sublime and Beautiful*, ed. James T. Boulton (Notre Dame: University of Notre Dame Press, 1968).

6. Alan Richardson, *The Neural Sublime: Cognitive Theories and Romantic Texts* (Baltimore: Johns Hopkins University Press, 2010), chapter 2.

7. Tom Huhn, *Imitation and Society: The Persistence of Mimesis in the Aesthetics of Burke, Hogarth, and Kant* (University Park: Pennsylvania State University Press, 2004), 62, 66.

8. William Hogarth, *The Analysis of Beauty*, ed. Ronald Paulson (New Haven: Yale University Press, 1997), 34, 33. When Hogarth first introduces these lines of grace and beauty in the chapter on "Intricacy," he offers two paradigmatic experiences as examples of the perceptual processes involved in tracing winding lines. The first is the experience of reading the letters of the alphabet, which, according to Hogarth, is physiologically successive and yet gives the satisfying illusion of taking it all in "on one sudden view" by virtue of "the amazing ease and swiftness, with which it [the eye] performs this task" (33). The second is the experience of watching a country dancer by which the observer is physically stationary and yet has the satisfying illusion of movement. Hogarth seems to be getting at how something may be lively—he finds ideal beauty in the human body—and yet still recognizable as an object, moving and changing yet remaining itself. The dancer—like the winding line our eye follows as it reads—is "bewitching to the sight" (34) both in the sense of captivating and commanding vision's attention and in the sense of making it seem as though something that isn't really there—there is no real connecting line—can be seen.

9. See W. J. T. Mitchell's "Eye and Ear: Edmund Burke and the Politics of Sensibility" in *Iconology: Image, Text, Ideology* (Chicago: University of Chicago Press, 1986). Alternatively, Geraldine Friedman offers a deconstructive explanation, whereby the linguistic primacy of the last part of the *Enquiry* underlines the paradoxical terms in which Burke exemplifies his sensationist model of the sublime all along the treatise. Burke's itinerary from the Introduction on Taste to Part V, Friedman argues, betrays the extent to which sensation rides on the symbolic, while language gains its value from feeling. "Thus word and image exist in a strange relation of continuity and rupture; for it is by denying

sensationism that language, with its superior expressivity, maximizes sensation, in the sense of emotional effect. From this, we are forced to conclude that Part V is both the point to which the *Enquiry* tends and the point at which it self-destructs, because Burke's sensationism culminates in language precisely to the extent that it also dies there" (*The Insistence of History: Revolution in Burke, Wordsworth, Keats, and Baudelaire* [Stanford: Stanford University Press, 1996], 25). As my discussions of musical rhythm and mediation in the previous chapters suggest, and as I argue here, I do not believe language and sensation to be deconstructively antithetical.

10. We saw in the previous chapter that Sterne develops a similar critique of Locke's philosophy of language. Burke comments that "It seems to be an odd subject of dispute with any man, whether he has ideas in his mind or not . . . But strange as it may appear, we are often at a loss to know what ideas we have of things, or whether we have any ideas at all upon some subjects" (168). Burke here seems proleptically to invoke a Toby Shandy, who makes two hours feel an age by speaking of Locke's analysis of time without having any idea of what he talks about.

11. Eighteenth-century philosophers most often tackled the question by commenting on what has come to be known as "Molyneux's problem." In a question posed to Locke, the Dublin lawyer William Molyneux asks whether a blind man whose use of his eyes was suddenly restored could correctly name shapes he had known only by touch. Committed to avoiding innate ideas—ones that in this case could translate the abstractions generated by many occasions of tactile experience into an abstraction that can apply to the first occasion of ocular experience—Locke agrees with Molyneux that the answer must be no. But if subsequent discussions uphold this bottom line, some contest Locke's privileging of the data of vision over that of touch in his explanation. Moreover, debates about how the data of different sensory modalities might or might not be comparable often consider temporality and liveliness, which turn out to be key for Burke. For helpful discussions of the Molyneux problem, see Marjolein Degenaar, *Molyneux's Problem: Three Centuries of Discussion on the Perception of Forms*, trans. Michael J. Collins (Dordrecht: Kluwer Academic Publishing, 1996); Michael J. Morgan, *Molyneux's Question: Vision, Touch, and the Philosophy of Perception* (Cambridge: Cambridge University Press, 1977); and Kevis Goodman, *Georgic Modernity and British Romanticism: Poetry and the Mediation of History* (Cambridge: Cambridge University Press, 2004), 50–54.

12. Denis Diderot, "Letter on the Blind" (1749), in *Diderot's Early Philosophical Works* trans. and ed. Margaret Jourdain (Chicago: Open Court, 1916; rpt. London: Forgotten Books, 2012), 68–141. In weighing in on Molyneux's problem in his "Letter on the Blind," Diderot argues, among other things, that touch and sight are of equal importance and reciprocally aid one another; but he also emphasizes that neither one depends on the other. Additionally, he argues that because sensations are different from the objects that occasion them it takes training and time for each sense modality to effectively identify what it senses. This suggests that the ideas our perceptions generate may be misleading with regards to the "real truth" about phenomena. Diderot often returns to such skeptical arguments along his discussion, but does so with a doubleness that alerts us to their truthfulness as well as to

their pragmatic shortcomings. And yet in his extensive discussion of Saunderson in the same letter, Diderot suggests without irony that the ideas generated from vision may be misleading in a way that the ideas generated by touch are not.

13. Ann Radcliffe, *The Mysteries of Udolpho: A Romance*, ed. and intro. Jacqueline Howard (London: Penguin, 2001).

14. Miller continues: "What made evening special to Collins and others, however, was that its liminality crystallized an idea of the ephemeral, the peculiarity of a momentary lyric utterance" (14).

15. Ann Wierda Rowland, "Romantic Poetry and the Romantic Novel" in *The Cambridge Companion to British Romantic Poetry*, ed. James Chandler and Maureen N. McLane (Cambridge: Cambridge University Press, 2008), 132.

16. Ingrid Horrocks argues that in *The Mysteries of Udolpho* viewing landscape, like reading literature and composing poetry, is a form of relief, thus putting these activities on par. When Emily can't read, she takes solace in looking out her window; this is a form of evacuation of self insofar as it "stage[s] the blocking and enabling of sympathetic communication through quotation and poetry" ("'Her Ideas Arranged Themselves': Re-membering Poetry in Radcliffe," *Studies in Romanticism* 47.4 [2008]: 509). Horrocks observes that for Radcliffe such opening up of individual consciousness entails less of a deepening of subjectivity than an exteriorization of the self. Similarly, G. Gabrielle Starr's study of relations between verse and novels in the eighteenth century highlights the preoccupation of both genres with problems of community and communicability. See *Lyric Generations: Poetry and the Novel in the Long Eighteenth Century* (Baltimore: Johns Hopkins University Press, 2004).

17. Marshall Brown, *The Gothic Text* (Stanford: Stanford University Press, 2005), 168. But if the medium is essentially flat, it is also essentially temporal, and it is such temporalization that is key for Radcliffe's character construction. I develop this argument in the next section of this chapter. Brown also complains that the landscape scenes convey a failure of vision on Radcliffe's part, which he means pejoratively, but I believe to be precisely part of Radcliffe's program (in the Burkean sense of the advantages of blindness). But even as Brown seems to me to miss what is at stake for Radcliffe, he does offer some very important observations about the gothic genre more generally. He recognizes that the gothic probes "the limbo of spacelessness and timelessness where the soul is most intimately tested," and casts this limbo as "a shadow realm that effaces all distinctions." Gothic heroes and heroines placed in such extreme conditions "discover . . . the essential activity of the soul, which is the production of time" (136, 137, 139). For more about Radcliffe's landscape scenes, see Malcolm Ware, who argues that Radcliffe's "scenery is described in such great detail and with so very few personal, individualized reactions involving the characters that the action is halted and the scenes contemplated much as one would a painting" ("The Telescope Reversed: Ann Radcliffe and Natural Scenery" in *A Provision of Human Nature: Essays on Fielding and Others in Honor of Miriam Austin Locke* [Tuscaloosa: University of Alabama Press, 1977], 174).

18. Eve Kosofsky Sedgwick, "The Character in the Veil: Imagery of the Surface in the Gothic Novel" *PMLA* 96.2 (1981): 255–70; see especially 264–65.

19. Mark Evan Bonds, *Wordless Rhetoric: Musical Form and the Metaphor of the Oration* (Cambridge: Harvard University Press, 1991), 53. Bonds argues that the metaphor of oration dominates eighteenth-century discussions, whereas nineteenth-century discussions emphasize organism. Oration focuses on listener, compositional unfolding and, hence, temporality. Organism, by contrast, stresses spatial dimensions.

20. Though this essay was published posthumously, Smith worked on it for a long time—starting in 1778 and continuing with interruptions to the last days of his life. And while he asked that his notes and manuscripts be destroyed after his death, he spared this essay along with a handful of others. For the composition history of this essay, see Neil De Marchi, "Smith on Ingenuity, Pleasure, and the Imitative Arts" in *The Cambridge Companion to Adam Smith,* ed. Knud Haakonssen (Cambridge: Cambridge University Press, 2006), 136–57. Radcliffe could not have read the essay before publishing *The Mysteries of Udolpho*, but as Bonds' study indicates, the conceptualization that Smith for many years worked to articulate with such clarity in this essay would have been available for a while among those interested in musical form.

21. Adam Smith, "Of the Nature of That Imitation Which Takes Place in What Are Called the Imitative Arts" in *Essays on Philosophical Subjects* (London, 1795).

22. Contra Rameau, who privileges harmony and, thus, synchronic resonances of sound, Rousseau privileges melody and successive relations. Thus Rousseau, very much like Smith, sees patterned succession as the most important element of music. But while Rousseau identifies this element as mimesis, Smith does not. I suspect that Smith downplays how much he agrees with Rousseau because he is committed to a sharper differentiation between intellectual and sentimental pleasures. Rousseau lodges music's moral value solely in its evocation of feelings; in his account, melody imitates voice in its expression of passions, signifies our affections, and thus arouses emotion. See Rousseau's *Essay on the Origin of Languages*. Smith, we will soon see, contends that even more important than sentiment, music instantiates thinking.

23. Maria Semi reads these passages as Smith's putting his aesthetics in dialogue with Hume's theory of causality; see *Music as a Science of Mankind in Eighteenth-Century Britain,* trans. Timothy Keates (Farnum, UK: Ashgate, 2012), 100–102. While the connection with Hume seems to me helpful, I would underline his other two associative principles—contiguity and similarity—as most relevant to music's case.

24. De Marchi, "Smith on Ingenuity, Pleasure, and the Imitative Arts," 143.

25. Frances Ferguson, *Solitude and the Sublime: Romanticism and the Aesthetics of Individuation* (New York: Routledge, 1992), 2.

26. Terry Castle, "The Spectralization of the Other in *The Mysteries of Udolpho*" in *The Female Thermometer: Eighteenth-Century Culture and the Invention of the Uncanny* (Oxford: Oxford University Press, 1995),123.

27. Adela Pinch, "Phantom Feelings: Emotional Occupation in *The Mysteries of Udolpho*" in *Strange Fits of Passion: Epistemologies of Emotion, Hume to Austen* (Stanford: Stanford University Press, 1996), 116. Pinch takes this move from imagining to image as one of Radcliffe's strategies for obscuring the sources of emotion. Unlike Castle, however,

for Pinch this does not indicate individualist self-absorption, but rather the possibility that feelings are constitutively intersubjective.

28. William Wordsworth, *The Prelude: 1799, 1805, 1850*, ed. Jonathan Wordsworth (New York: W. W. Norton, 1979), Book Tenth (1805), lines 57, 60.

29. Duncan also notes the importance of dynamics of "prospect, retrospect, and present scene" (43) in this episode, but for different purposes—purposes that highlight content and ideological concerns. Interested in the sexual politics that Radcliffe's romance revises, Duncan focuses on the role of St Aubert as "the benevolent and rationalist father" (43) whose "power relies on a peculiar negativity of presence" (43–44).

30. Alex Woloch, *The One vs. the Many: Minor Characters and the Space of the Protagonist in the Novel* (Princeton: Princeton University Press, 2003).

31. Or, in the case of a critical stance, to ask how much value we should place on it. Think of Dickens's rendering of laws that can neither recognize nor benefit from Magwitch's reform, or Wilde's rendering of aesthetics that foster the illusion of displacing continuity onto art. Dorian Gray, from this perspective, is the horrific materialization of Laurentini's hope.

32. Castle identifies in *The Mysteries of Udolpho* Romantic self-absorption, arguing that this novel is symptomatic of "the new sensibility of the late eighteenth century [which] was, quite literally, a growing sense of the ghostliness of other people" (125).

33. Walpole explains in the first paragraph of the first preface to *The Castle of Otranto* that the genesis of the work is as follows: "The following work was found in the library of an ancient Catholic family in the north of England. It was printed at Naples, in the black letter, in the year 1529. How much sooner it was written does not appear. The principal incidents are such as were believed in the darkest ages of christianity; but the language and conduct have nothing that savours of barbarism. The style is the purest Italian. If the story was written near the time when it is supposed to have happened, it must have been between 1095, the era of the first crusade, and 1243, the date of the last, or not long afterwards. There is no other circumstance in the work, that can lead us to guess at the period in which the scene is laid. The names of the actors are evidently fictitious, and probably disguised on purpose: yet the Spanish names of the domestics seem to indicate that this work was not composed until the establishment of the Arragonian kings in Naples had made Spanish appellations familiar in that country. The beauty of the diction, and the zeal of the author (moderated, however, by singular judgment), concur to make me think, that the date of the composition was little antecedent to that of the impression. Letters were then in their most flourishing state in Italy, and contributed to dispel the empire of superstition, at that time so forcibly attacked by the reformers. It is not unlikely, that an artful priest might endeavour to turn their own arms on the innovators; and might avail himself of his abilities as an author to confirm the populace in their ancient errors and superstitions. If this was his view, he has certainly acted with signal address. Such a work as the following would enslave a hundred vulgar minds, beyond half the books of controversy that have been written from the days of Luther to the present hour" (*The Castle of Otranto* [New York: Dover, 1966], 17).

34. Ruth Mack writes: "What Walpole sees, long before the new historicists do, is that one destination of wonder is historiographic practice itself, in its continued entertainment of the possibility that writing in the present can offer us access to the past's 'real presence.'" ("Horace Walpole and the Objects of Literary History," *ELH* 75.2 (2008): 367–87, 382). In *Literary Historicity: Literature and Historical Experience in Eighteenth-Century Britain* (Stanford: Stanford University Press, 2009), Mack argues more generally that mid-century novelists raise profound questions about historical consciousness. Writers of the midcentury "are less interested in proving that an accurate description of the past is possible or impossible than they are in analyzing their experience of the past, especially as it remains part of their current world" (19).

35. Sean Silver makes the case for empiricist associationism as modeling the mind's labors through the two principles of well-organized encasement and materialist digression. His models of encasement are predominantly spatial—museum, library, laboratory; and in some brief comments about this model's conceptualization of time, Silver emphasizes spatializing and quantifying approaches in Hooke's and Locke's discussions. Silver reads Walpole's *The Mysterious Mother* as the final exhibit in a case about inwardness, showing that if the body is discursive—public—this arises from moments in which we glimpse a deeply interiorized—at least spatially concealed—embodiment. See *The Mind Is a Collection: Case Studies in Eighteenth-Century Thought* (Philadelphia: University of Pennsylvania Press, 2015).

36. In this sense, Radcliffe is as indifferent to the aims of the bildungsroman that becomes increasingly influential at the end of the eighteenth century as she is to the aims of historicism and the historical novel. She may focus on private experience—on "ordinary administration: a time of 'lived experience' and individual growth—a time filled with 'opportunities', but which excludes by definition both the crisis and genesis of a culture" (12), and on the task of "stringing together 'experiences'" (48), as Franco Moretti defines the bildung—but these can hardly be identified in her novels as servicing something like the evolving construction of a "personality"—that unique individual whose achievement the bildungsroman aims for. See Moretti's *The Way of the World: The Bildungsroman in European Culture* (London: Verso, 1987).

37. Deidre Shauna Lynch, *The Economy of Character: Novels, Market Culture, and the Business of Inner Meaning* (Chicago: University of Chicago Press, 1998).

38. Jane Austen, *Sense and Sensibility*, ed. and intro. Ros Ballaster (New York: Penguin Books, 1995).

39. Nicholas Dames, *Amnesiac Selves: Nostalgia, Forgetting, and British Fiction, 1810–1870* (Oxford: Oxford University Press, 2001).

40. Dames studies nineteenth-century associationism, which he understands as aiming to causally emplot the mind. As I have been arguing throughout this book, the sensibility chronotope underlines metonymy and contiguity more than causality as key principles of association, and as such conceives of the mind's composition more by the model of music than of emplotment.

41. Dames, *The Physiology of the Novel: Reading, Neural Science, and the Form of Victorian Fiction* (Oxford: Oxford University Press, 2007), 10–11.

CODA

1. Jonathan Crary, *24/7: Late Capitalism and the Ends of Sleep* (London: Verso, 2013), 3, 8, 9.

2. Meghan O'Rourke, "Is 'The Clock' Worth the Time?" *New Yorker*, July 18, 2012.

3. Roberta Smith, "As in Life, Timing Is Everything in the Movies," *New York Times*, February, 3, 2011.

4. Peter Bradshaw, "Christian Marclay's The Clock: A Masterpiece of Our Times," *Guardian* April 7, 2011.

5. On the scale of fine-art exhibitions, this work draws large crowds around the world, with people willing to wait their turn at any hour of day or night, rain or shine. See Randy Kennedy's report for the *New York Times* "Flock Around the Clock," February 16, 2011. "Even when they involve 14-foot tiger sharks preserved in formaldehyde or pictures of Jeff Koons in flagrante delicto, commercial art gallery exhibitions in New York don't often draw capacity crowds. And they almost never move people to line up along the postindustrial streets of Chelsea on weekday winter mornings as if a sample sale were under way. But at 9:30 Wednesday morning, with the cold wind raising tears in their eyes, Nick and Elspeth Macdonald from Park Slope, Brooklyn, huddled dutifully in front of the Paula Cooper Gallery along with 20 other people. . . . Over the last three and a half weeks the exhibition has built itself into an unlikely kind of rock-concert phenomenon, with crowds lined up on West 21st Street as late at 2 a.m. on Saturdays, when the gallery remains open overnight to show the film in its entirety, in a makeshift theater space that seats about 80 people. And as the exhibition approaches its final weekend (it closes on Saturday), the crowds have continued to build. By the gallery's rough count, more than 780 people passed through on Tuesday alone—some spending only a few minutes, some spending hours—and almost 7,000 people have watched the piece since gallery workers began to keep track on Feb. 4." *The Clock* has drawn such crowds in two separate showings in New York, as well as in London, Jerusalem, Boston, Minneapolis, and many other cities.

6. Garrett Stewart, *Framed Time: Towards a Postfilmic Cinema* (Chicago: University of Chicago Press, 2007), 114. Stewart focuses in *Framed Time* on cinema's transitioning from film to pixel, a change in medium that, he contends, enables a new approach to time—from "segmental transit" to "fragmental transformation" (6)—as well as establishes the stakes for his narratography, which I discussed in Chapter 3.

7. Sergei Eisenstein in "Dickens, Griffith, and the Film Today" is the first to relate cinematic and novelistic techniques, but he considers only visualization and plotting. Stewart discovers "a whole new mode of kinetic sequencing in which discrepant juxtaposition is submitted to continual resynthesis" (*Framed Time*, 250). Such logic of montage at

the register of syntax can construct novelistic temporalities that are radically different from the chronologies of emplotment for which high realism is notorious, an argument Stewart elaborates in his subsequent narratographic study of Victorian fiction, which I discuss in Chapter 3.

8. James Chandler, *An Archaeology of Sympathy: The Sentimental Mode in Literature and Cinema* (Chicago: University of Chicago Press, 2013).

# *Bibliography*


Abercrombie, David. *Studies in Phonetics and Linguistics.* London: Oxford University Press, 1965.

Abrams, M. H. *Natural Supernaturalism: Tradition and Revolution in Romantic Literature.* New York: W. W. Norton, 1971.

Addison, Joseph and Richard Steele. *The Spectator.* Ed. Donald F. Bond. 5 vols. Oxford: Clarendon Press, 1965.

Agamben, Giorgio. *The Time that Remains: A Commentary on the Letter to the Romans.* Trans. Patricia Dailey. Stanford: Stanford University Press, 2005.

———. *Infancy and History: On the Destruction of Experience.* Trans. Liz Heron. London: Verso, 2007.

Alkon, Paul K. *Defoe and Fictional Time.* Athens: University of Georgia Press, 1979.

———. "Joshua Steele and the Melody of Speech." *Language and Speech* 2, no. 3 (1959): 154–174.

Anderson, Benedict. *Imagined Communities: Reflections on the Origin and Spread of Nationalism.* London: Verso, 1983.

Aravamudan, Srinivas. *Enlightenment Orientalism: Resisting the Rise of the Novel.* Chicago: University of Chicago Press, 2012.

Augustine. *Confessions.* Trans. Albert C. Outler. Rev. trans. Mark Vessey. New York: Barnes & Noble Books, 2007.

Austen, Jane. *Sense and Sensibility.* Ed. and intro. Ros Ballaster. New York: Penguin Books, 1995.

Bach, Johann S., Glenn Gould, and Tim Page. *A State of Wonder: The Complete Goldberg Variations 1955 & 1981.* Sony, 2002.

Baier, Annette C. *A Progress of Sentiments: Reflections on Hume's* Treatise. Cambridge, MA: Harvard University Press, 1991.

Bakhtin, M. M. *The Dialogic Imagination: Four Essays.* Ed. Michael Holquist. Trans. Caryl Emerson and Michael Holquist. Austin: University of Texas Press, 1981.

Barchas, Janine. *Graphic Design, Print Culture, and the Eighteenth-Century Novel.* Cambridge: Cambridge University Press, 2003.

Barker-Benfield, G. J. *The Culture of Sensibility: Sex and Society in Eighteenth-Century Britain.* Chicago: University of Chicago Press, 1992.

———. "Sensibility." In *An Oxford Companion to the Romantic Age: British Culture, 1776–1832,* ed. Iain McCalman, 102–14. Oxford: Oxford University Press, 1999.

Beck, Lewis White. "World Enough, and Time." In *Probability, Time, and Space in Eighteenth-Century Literature*, ed. Paula R. Backscheider, 113–40. New York: AMS Press, 1979.

Bentham, Jeremy. *The Principles of Morals and Legislation*. London, 1780. Rpt. Amherst, NY: Prometheus Books, 1988.

Bergson, Henri. *Time and Free Will: An Essay on the Immediate Data of Consciousness*. Trans. F. L. Pogson. Dover, 2001.

Berkeley, George. *A Treatise Concerning the Principles of Human Knowledge*. Dublin, 1710.

Blattner, William. "Temporality." In *A Companion to Heidegger*, ed. Hubert L. Dreyfus and Mark A. Wrathall. 311–34. Malden, MA: Blackwell, 2005.

Bonds, Mark Evan. *Wordless Rhetoric: Musical Form and the Metaphor of the Oration*. Cambridge, MA: Harvard University Press, 1991.

Bradshaw, Peter. "Christian Marclay's The Clock: A Masterpiece of Our Times." *Guardian*. April 7, 2011. https://www.theguardian.com/film/filmblog/2011/apr/07/christian-marclay-the-clock.

Brissenden, R. F. *Virtue in Distress: Studies in the Novel of Sentiment from Richardson to Sade*. London: Macmillan, 1974.

Brown, Marshall. *The Gothic Text*. Stanford: Stanford University Press, 2005.

Brown, Tony C. *The Primitive, the Aesthetic, and the Savage: An Enlightenment Problematic*. Minneapolis: University of Minnesota Press, 2012.

Burke, Edmund. *A Philosophical Enquiry into the Origin of Our Ideas of the Sublime and Beautiful*. Ed. James T. Boulton. Notre Dame: University of Notre Dame Press, 1968.

Campbell, T. D. "Francis Hutcheson: 'Father' of the Scottish Enlightenment." In *The Origins and Nature of the Scottish Enlightenment*. Ed. R. H. Campbell and Andrew S. Skinner, 167–85. Edinburgh: John Donald, 1982.

Carey, Daniel. *Locke, Shaftesbury, and Hutcheson: Contesting Diversity in the Enlightenment and Beyond*. Cambridge: Cambridge University Press, 2006.

Carlson, Julia S. *Romantic Marks and Measures: Wordsworth's Poetry in Fields of Print*. Philadelphia: University of Pennsylvania Press, 2016.

Cash, Arthur H. *Laurence Sterne: The Later Years*. London: Methuen, 1986.

Castle, Terry. "The Spectralization of the Other in *The Mysteries of Udolpho*." In *The Female Thermometer: Eighteenth-Century Culture and the Invention of the Uncanny*, 120–39. New York: Oxford University Press, 1995.

Chambers, Ephraim. *Cyclopaedia, or, An Universal Dictionary of Arts and Sciences*. London, 1728.

Chandler, James. *An Archaeology of Sympathy: The Sentimental Mode in Literature and Cinema*. Chicago: University of Chicago Press, 2013.

Chartier, Roger. *Inscription and Erasure: Literature and Written Culture from the Eleventh to the Eighteenth Century*. Trans. Arthur Goldhammer. Philadelphia: University of Pennsylvania Press, 2007.

Cheyne, George. *The English Malady*. London, 1733. Rpt. Delmar, NY: Scholars' Facsimiles and Reprints, 1976.

Clune, Michael W. *Writing Against Time*. Stanford: Stanford University Press, 2013.

Cohen, Murray. *Sensible Words: Linguistic Practice in England, 1640–1785*. Baltimore: Johns Hopkins University Press, 1977.

Crary, Jonathan. *24/7: Late Capitalism and the Ends of Sleep*. London: Verso, 2013.

Dames, Nicholas. *Amnesiac Selves: Nostalgia, Forgetting, and British Fiction, 1810–1870*. Oxford: Oxford University Press, 2001.

———. *The Physiology of the Novel: Reading, Neural Science, and the Form of Victorian Fiction*. Oxford: Oxford University Press, 2007.

Darwall, Stephen. *The British Moralists and the Internal "Ought": 1640–1740*. Cambridge: Cambridge University Press, 1995.

De Bolla, Peter. *The Discourse of the Sublime: Readings in History, Aesthetics, and the Subject*. Oxford: Basil Blackwell, 1989.

Defoe, Daniel. *Robinson Crusoe*. Ed. Michael Shinagel. 2nd ed. New York: W. W. Norton, 1994.

Degenaar, Marjolein. *Molyneux's Problem: Three Centuries of Discussion on the Perception of Forms*. Trans. Michael J. Collins. Dordrecht: Kluwer Academic Publishing, 1996.

Deleuze, Gilles. *Empiricism and Subjectivity: An Essay on Hume's Theory of Human Nature*. Trans. Constantin V. Boundas. New York: Columbia University Press, 1991.

De Marchi, Neil. "Smith on Ingenuity, Pleasure, and the Imitative Arts." In *The Cambridge Companion to Adam Smith*, ed. Knud Haakonssen, 136–57. Cambridge: Cambridge University Press, 2006.

De Warren, Nicolas. "Time" in *The Routledge Companion to Phenomenology*, ed. Sebastian Luft and Soren Overgaard, 190–201. Routledge, 2012.

Diderot, Denis. "In Praise of Richardson." In *Diderot: Selected Writings on Art and Literature*. Trans and ed. Geoffrey Bremner, 82–97. London: Penguin Books, 1994.

———. "Letter on the Blind." In *Diderot's Early Philosophical Works*. Trans. and ed. Margaret Jourdain, 68–141. Chicago: Open Court, 1916. Rpt. London: Forgotten Books, 2012.

Duncan, Ian. *Modern Romance and the Transformations of the Novel: The Gothic, Scott, Dickens*. Cambridge: Cambridge University Press, 1992.

Elfenbein, Andrew. *Romanticism and the Rise of English*. Stanford: Stanford University Press, 2009.

Ermarth, Elizabeth Deeds. *Realism and Consensus in the English Novel: Time, Space, and Narrative*. Princeton: Princeton University Press, 1983.

Ferguson, Frances. *Solitude and the Sublime: Romanticism and the Aesthetics of Individuation*. New York: Routledge, 1992.

Festa, Lynn. *Sentimental Figures of Empire in Eighteenth-Century Britain and France*. Baltimore: Johns Hopkins University Press, 2006.

Fliegelman, Jay. *Declaring Independence: Jefferson, Natural Language, and the Culture of Performance*. Stanford: Stanford University Press, 1993.

Foucault, Michel. *Discipline and Punish: The Birth of the Prison*. Trans. Alan Sheridan. Harmondsworth: Penguin, 1977.

Freedman, William. *Laurence Sterne and the Origins of the Musical Novel.* Athens: University of Georgia Press, 1978.

Fried, Michael. *Absorption and Theatricality: Painting and Beholder in the Age of Diderot.* Chicago: University of Chicago Press, 1980.

Friedman, Geraldine. *The Insistence of History: Revolution in Burke, Wordsworth, Keats, and Baudelaire.* Stanford: Stanford University Press, 1996.

Gallagher, Catherine. "Formalism and Time." In *Reading for Form*, ed. Susan Wolfson and Marshall Brown, 305–27. Seattle: University of Washington Press, 2006.

Genette, Gerard. *Narrative Discourse: An Essay in Method.* Trans. Jane E. Lewin. Ithaca: Cornell University Press, 1980.

———. *Narrative Discourse Revisited.* Trans. Jane E. Lewin. Ithaca: Cornell University Press, 1988.

Goodman, Kevis. *Georgic Modernity and British Romanticism: Poetry and the Mediation of History.* Cambridge: Cambridge University Press, 2004.

Gordon, Scott Paul. *The Power of the Passive Self in English Literature, 1640–1770.* Cambridge: Cambridge University Press, 2002.

Goring, Paul. *The Rhetoric of Sensibility in Eighteenth-Century Culture.* Cambridge: Cambridge University Press, 2005.

Hartley, David. *Observations on Man, His Frame, His Duty, and His Expectations.* London, 1749.

Heidegger, Martin. *Being and Time.* Trans. John Macquarrie and Edward Robinson. Oxford: Blackwell, 1962.

Herries, John. *The Elements of Speech.* London, 1773. Rpt. Menston: Scolar Press, 1968.

Hogarth, William. *The Analysis of Beauty.* Ed. Ronald Paulson. New Haven: Yale University Press, 1997.

Horrocks, Ingrid. "'Her Ideas Arranged Themselves': Re-membering Poetry in Radcliffe." *Studies in Romanticism* 47, no. 4 (2008): 507–27.

Howell, Wilbur Samuel. *Eighteenth-Century British Logic and Rhetoric.* Princeton: Princeton University Press, 1971.

Hoy, David Couzens. *The Time of Our Lives: A Critical History of Temporality.* Cambridge, MA: MIT Press, 2009.

Hudson, Nicholas. *Writing and European Thought, 1600–1830.* Cambridge: Cambridge University Press, 1994.

Huhn, Tom. *Imitation and Society: The Persistence of Mimesis in the Aesthetics of Burke, Hogarth, and Kant.* University Park: Pennsylvania State University Press, 2004.

Hume, David. *An Abstract of a Book Lately Published Entitled* A Treatise of Human Nature. London, 1740.

———. *A Treatise of Human Nature.* Buffalo: Prometheus Books, 1992.

Husserl, Edmund. *On The Phenomenology of the Consciousness of Internal Time (1893–1917)* Trans. John Barnett Brough, vol. 4 of *Edmund Husserl, Collected Works*, ed. Rudolf Bernet. Dordrecht: Kluwer Academic Publishers, 1991.

Hutcheson, Francis. *An Essay on the Nature and Conduct of the Passions and Affections, with*

*Illustrations on the Moral Sense*. Ed. and intro. Aaron Garrett. Indianapolis: Liberty Fund, 2002.

———. *An Inquiry into the Original of Our Ideas of Beauty and Virtue in Two Treatises*, ed. and intro. Wolfgang Leidhold, rev. ed. Indianapolis: Liberty Fund, 2008.

Johnson, Samuel. *A Dictionary of the English Language*. 2nd ed. London, 1755.

Kallich, Martin. *The Association of Ideas and Critical Theory in Eighteenth-Century England: A History of a Psychological Method in English Criticism*. The Hague: Mouton, 1970.

Kassler, Jamie C. *Inner Music: Hobbes, Hooke and North on Internal Character*. London: Athlone, 1995.

———. "Representing Speech Through Musical Notation." *Journal of Musicological Research* 24, nos. 3–4 (2005): 227–39.

———. *The Science of Music in Britain, 1714–1830: A Catalogue of Writings, Lectures and Inventions*. 2 vols. New York: Garland, 1979.

Kavanagh, Thomas M. *Esthetics of the Moment: Literature and Art in the French Enlightenment*. Philadelphia: University of Pennsylvania Press, 1996.

Kennedy, Randy. "Flock Around 'The Clock.'" *New York Times*. 16 February 2011. http://www.nytimes.com/2011/02/17/arts/design/17christianmarclay.html.

Kermode, Frank. *The Sense of an Ending: Studies in the Theory of Fiction with a New Epilogue*. Oxford: Oxford University Press, 2000.

Ketcham, Michael G. *Transparent Designs: Reading, Performance, and Form in* The Spectator *Papers*. Athens: University of Georgia Press, 1985.

Keymer, Thomas. *Richardson's* Clarissa *and the Eighteenth-Century Reader*. Cambridge: Cambridge University Press, 1992.

———. *Sterne, the Moderns, and the Novel*. Oxford: Oxford University Press, 2002.

King, Amy M. "Dilatory Description and the Pleasures of Accumulation: Toward a History of Novelistic Length." In *Narrative Middles: Navigating the Nineteenth-Century British Novel*, ed. Caroline Levine and Mario Ortiz-Robles, 161–94. Columbus: Ohio State University Press, 2011.

Kinkead-Weekes, Mark. *Samuel Richardson: Dramatic Novelist*. Ithaca: Cornell University Press, 1973.

Kramnick, Jonathan. *Actions and Objects from Hobbes to Richardson*. Stanford: Stanford University Press, 2010.

Lamb, Jonathan. *Sterne's Fiction and the Double Principle*. Cambridge: Cambridge University Press, 1989.

Lehman, Benjamin H. "Of Time, Personality, and the Author." Rpt. in *Laurence Sterne: A Collection of Critical Essays*, ed. John Traugott, 21–33. Englewood, NJ: Prentice-Hall, 1968.

Leslie, John C. "Music's Sentimental Role in *Tristram Shandy*." *Papers on Language and Literature* 41, no. 1 (2005): 55–66.

Levenson, Michael. "Reading Time." *Novel: A Forum on Fiction* 42, no. 3 (2009): 511–16.

Levine, Caroline. "An Anatomy of Suspense: The Pleasurable, Critical, Ethical, Erotic Middle of *The Woman in White*." In *Narrative Middles: Navigating the Nineteenth-Century*

*British Novel*, ed. Caroline Levine and Mario Ortiz-Robles, 195–214. Columbus: Ohio State University Press, 2011.

Locke, John. *An Essay Concerning Human Understanding*. Ed. and intro. Peter H. Nidditch. Oxford: Oxford University Press, 1975.

Luhmann, Niklas. *Art as a Social System*. Trans. Eva M. Knodt. Stanford: Stanford University Press, 2000.

———. *Love as Passion: The Codification of Intimacy*. Trans. Jeremy Gaines and Doris L. Jones. Cambridge: Polity, 1986. Rpt. Stanford: Stanford University Press, 1998.

Lynch, Deidre Shauna. *The Economy of Character: Novels, Market Culture, and the Business of Inner Meaning*. Chicago: University of Chicago Press, 1998.

———. "Austen Extended/Austen for Everyday Use," in *Imagining Selves: Essays in Honor of Patricia Meyer Spacks*, ed. Rivka Swenson and Elise Lauterbach. Newark: University of Delaware Press, 2008.

———. *Loving Literature: A Cultural History*. Chicago: University of Chicago Press, 2015.

MacIntosh, J. J. "Berkeley's Views on Time." In *New Essays on Rationalism and Empiricism*, ed. Charles E. Jarrett, John King-Farlow and F.J. Pelletier, 153–63. Guelph, Ont.: Canadian Association for Publishing in Philosophy, 1978.

Mack, Ruth. "Horace Walpole and the Objects of Literary History." *ELH* 75, no. 2 (2008): 367–87.

———. *Literary Historicity: Literature and Historical Experience in Eighteenth-Century Britain*. Stanford: Stanford University Press, 2009.

Macpherson, Sandra. *Harm's Way: Tragic Responsibility and the Novel Form*. Baltimore: Johns Hopkins University Press, 2010.

Marshall, David. *The Figure of Theater: Shaftesbury, Defoe, Adam Smith, and George Eliot*. New York: Columbia University Press, 1986.

———. *The Surprising Effects of Sympathy: Marivaux, Diderot, Rousseau, and Mary Shelley*. Chicago: University of Chicago Press, 1988.

Mayoux, Jean-Jacques. "Variations on the Time-Sense in *Tristram Shandy*." In *The Winged Skull*, ed. Arthur H. Cash and Joan M. Stedmond, 3–18. Kent, OH: Kent State University Press, 1971.

McKeon, Michael. *The Origins of the English Novel, 1660–1740*. Baltimore: Johns Hopkins University Press, 1987.

———. "Mediation as Primal Word: The Arts, the Sciences, and the Origins of the Aesthetic." In *This Is Enlightenment*, ed. Clifford Siskin and William Warner, 384–412. Chicago: University of Chicago Press, 2010.

Mendilow, A. A. *Time and the Novel*. 1952. Rpt. New York: Humanities Press, 1972.

Miller, Christopher R. *The Invention of Evening: Perception and Time in Romantic Poetry*. Cambridge: Cambridge University Press, 2006.

Mitchell, W. J. T. *Iconology: Image, Text, Ideology*. Chicago: University of Chicago Press, 1986.

Molesworth, Jesse. *Chance and the Eighteenth-Century Novel: Realism, Probability, Magic*. Cambridge: Cambridge University Press, 2010.

———. "Gothic Time, Sacred Time." *MLQ* 75, no. 1 (2014): 29–55.

Monboddo, James Burnett, Lord. *Of the Origin and Progress of Language*. London, 1773.

Moretti, Franco. "Serious Century." In *The Novel: History, Geography, and Culture*, ed. Franco Moretti, 2 vols., 1: 364–400. Princeton: Princeton University Press, 2006.

———. *The Way of the World: The Bildungsroman in European Culture*. London: Verso, 1987.

Morgan, Michael J. *Molyneux's Question: Vision, Touch, and the Philosophy of Perception*. Cambridge: Cambridge University Press, 1977.

Mullan, John. *Sentiment and Sociability: The Language of Feeling in the Eighteenth Century*. Oxford: Clarendon Press, 1988.

Nazar, Hina. *Enlightened Sentiments: Judgment and Autonomy in the Age of Sensibility*. New York: Fordham University Press, 2012.

North, Roger. *Roger North's Cursory Notes of Musicke (c. 1698–c. 1703); A Physical, Psychological, and Critical Theory*. Ed. Mary Chan and Jamie C. Kassler. Kensington, NSW: Unisearch, 1986.

———. *Roger North's The Musicall Grammarian, 1728*. Ed. with intro. and notes Mary Chan and Jamie C. Kassler. Cambridge: Cambridge University Press, 2006.

Odegard, Douglas. "Locke and the Specious Present." In *New Essays on Rationalism and Empiricism*, ed. Charles E. Jarrett, John King-Farlow, and F. J. Pelletier, 141–51. Guelph, Ont.: Canadian Association for Publishing in Philosophy, 1978.

O'Rourke, Meghan. "Is 'The Clock' Worth the Time?" *New Yorker*. July 18, 2012. http://www.newyorker.com/culture/culture-desk/is-the-clock-worth-the-time.

"Ought Sensibility to be Cherished, or Repressed?" *Monthly Magazine* 9 (1796): 706–9.

Pinch, Adela. *Strange Fits of Passion: Epistemologies of Emotion, Hume to Austen*. Stanford: Stanford University Press, 1996.

Plumb, J. H. "The Commercialization of Leisure in Eighteenth-Century England." In *The Birth of a Consumer Society: The Commercialization of Eighteenth-Century England*, ed. Neil McKendrick, John Brewer, and J. H. Plumb, 265–85. Bloomington: Indiana University Press, 1982.

Poulet, Georges. *Studies in Human Time*. Trans. Elliott Coleman. Baltimore: Johns Hopkins University Press, 1956.

Price, Leah. *The Anthology and the Rise of the Novel: From Richardson to George Eliot*. Cambridge: Cambridge University Press, 2000.

Radcliffe, Ann. *The Italian; or, the Confession of the Black Penitents: A Romance*. Ed. and intro. Robert Miles. London: Penguin Books, 2000.

———. *The Mysteries of Udolpho: A Romance*. Ed. and intro. Jacqueline Howard. London: Penguin Books, 2001.

Richardson, Alan. *The Neural Sublime: Cognitive Theories and Romantic Texts*. Baltimore: Johns Hopkins University Press, 2010.

Richardson, Samuel. *Clarissa, or, the History of a Young Lady*. Ed. and intro. Angus Ross. Harmondsworth, UK: Penguin Books, 1986.

———. *Pamela; or, Virtue Rewarded*. Ed. Thomas Keymer and Alice Wakely. Intro. Thomas Keymer. Oxford: Oxford University Press, 2001.

———. *Selected Letters*. Ed. and intro. John Carroll. Oxford: Oxford University Press, 1964.

Ricoeur, Paul. *Time and Narrative*. Trans. Kathleen McLaughlin and David Pellauer. 3 vols. Chicago: University of Chicago Press, 1984–1988.

Rimmon-Kenan, Shlomith. *Narrative Fiction: Contemporary Poetics*. London: Methuen, 1983.

Riskin, Jessica. *Science in the Age of Sensibility: The Sentimental Empiricists of the French Enlightenment*. Chicago: University of Chicago Press, 2002.

Rivers, Isabel. *Reason, Grace, and Sentiment: A Study of the Language of Religion and Ethics in England, 1660–1780*. 2 vols. Cambridge: Cambridge University Press, 2000.

Rousseau, G. S. "Discourses of the Nerve." In *Literature and Science as Modes of Expression*, ed. Frederick Amrine, 29–60. Dordrecht: Kluwer Academic Publishers, 1989.

———. "Nerves, Spirits, and Fibres: Towards Defining the Origins of Sensibility." In *Studies in the Eighteenth Century, III: Papers Presented at the Third David Nichol Smith Memorial Seminar, Canberra 1973*, ed. R. F. Brissenden and J. C. Eade, 137–57. Toronto: University of Toronto Press, 1976.

Rousseau, Jean-Jacques. "Essay on the Origin of Languages" in *The First and Second Discourses together with the Replies to Critics and Essay on the Origin of Languages*. Ed. and trans. Victor Gourevitch. New York: Harper and Row, 1986.

Sacks, Oliver. *Musicophilia: Tales of Music and the Brain*. Rev. ed. New York: Vintage Books, 2008.

Sedgwick, Eve Kosofsky. "The Character in the Veil: Imagery of the Surface in the Gothic Novel." *PMLA* 96, no. 2 (1981): 255–70.

Semi, Maria. *Music as a Science of Mankind in Eighteenth-Century Britain*. Trans. Timothy Keates. Farnham, UK: Ashgate, 2012.

Sheridan, Thomas. *A Course of Lectures on Elocution*. London, 1762. Ed. R. C. Alston. Menston, UK: Scolar Press, 1968.

Sherman, Stuart. *Telling Time: Clocks, Diaries, and English Diurnal Form, 1660–1785*. Chicago: University of Chicago Press, 1996.

Shklovsky, Viktor. "Art as Technique." Rpt. in *Modern Criticism and Theory: A Reader*, ed. David Lodge, 16–30. London: Longman, 1988.

Shortland, Michael. "Moving Speeches: Language and Elocution in Eighteenth-Century Britain." *History of European Ideas* 8, no. 6 (1987): 67–88.

Silver, Sean. *The Mind Is a Collection: Case Studies in Eighteenth-Century Thought*. Philadelphia: University of Pennsylvania Press, 2015.

Smith, Adam. "Of the Nature of that Imitation Which Takes Place in What Are Called the Imitative Arts." In *Essays on Philosophical Subjects*. London, 1795.

———. *The Theory of Moral Sentiments*. Ed. D. D. Raphael and A. L. Macfie. Oxford: Oxford University Press, 1976. Rpt. Indianapolis: Liberty Press, 1982.

Smith, Roberta. "As in Life, Timing Is Everything in the Movies." *New York Times*. February 3, 2011. http://www.nytimes.com/2011/02/04/arts/design/04marclay.html.

Soni, Vivasvan. "Committing Freedom: The Cultivation of Judgment in Rousseau's *Emile* and Austen's *Pride and Prejudice*." *The Eighteenth Century* 51, no. 3 (2010): 363–87.

———. *Mourning Happiness: Narrative and the Politics of Modernity*. Ithaca: Cornell University Press, 2010.

Spacks, Patricia Meyer. *Boredom: The Literary History of a State of Mind*. Chicago: University of Chicago Press, 1995.

Starr, G. Gabrielle. *Lyric Generations: Poetry and the Novel in the Long Eighteenth Century*. Baltimore: Johns Hopkins University Press, 2004.

Steele, Joshua. *An Essay Towards Establishing the Melody and Measure of Speech*. London, 1775. Rpt. Menston: Scolar Press, 1969.

Sterne, Laurence. *The Life and Opinions of Tristram Shandy, Gentleman*. Ed. Melvyn New and Joan New. Gainesville: University Press of Florida, 1978–84. Vols. 1–3 of Florida Edition of the Works of Laurence Sterne.

———. *Letters of Laurence Sterne*. Ed. Lewis Perry Curtis. Oxford: Clarendon Press, 1935.

Stewart, Garrett. *Framed Time: Toward a Postfilmic Cinema*. Chicago: University of Chicago Press, 2007.

———. *Novel Violence: A Narratography of Victorian Fiction*. Chicago: University of Chicago Press, 2009.

Tadié, Alexis. *Sterne's Whimsical Theatres of Language: Orality, Gesture, Literacy*. Aldershot, UK: Ashgate, 2003.

Thompson, E. P. *Customs in Common: Studies in Traditional Popular Culture*. London: Merlin Press, 1991.

Traugott, John. *Tristram Shandy's World: Sterne's Philosophical Rhetoric*. Berkeley: University of California Press, 1954.

Turetzky, Philip. *Time*. London: Routledge, 1998.

Van Sant, Ann Jessie. *Eighteenth-Century Sensibility and the Novel: The Senses in Social Context*. Cambridge: Cambridge University Press, 1993.

Walker, John. *A Rhetorical Grammar*. London, 1785. Rpt. Menston, UK: Scolar Press, 1972.

Walpole, Horace. *The Castle of Otranto*. New York: Dover, 1966.

Ware, Malcolm. "The Telescope Reversed: Ann Radcliffe and Natural Scenery." In *A Provision of Human Nature: Essays on Fielding and Others in Honor of Miriam Austin Locke*, ed. Donald Kay, 169–89. Tuscaloosa: University of Alabama Press, 1977.

Warner, William B. *Licensing Entertainment: The Elevation of Novel Reading in Britain, 1684–1750*. Berkeley: University of California Press, 1998.

———. "Proposal and Habitation: The Temporality and Authority of Interpretation in and about a Scene of Richardson's *Clarissa*." *Boundary 2*, 7, no. 2 (1979): 169–200.

Watt, Ian. *The Rise of the Novel: Studies in Defoe, Richardson, and Fielding*. Berkeley: University of California Press, 1957.

Woloch, Alex. *The One vs. the Many: Minor Characters and the Space of the Protagonist in the Novel*. Princeton: Princeton University Press, 2003.

Wordsworth, William. *Lyrical Ballads*. Ed. Michael Mason. 2nd ed. Harlow, UK: Pearson Longman, 2007.

Yahav-Brown, Amit. "Reasonableness and Domestic Fiction." *ELH* 73, no. 4 (2006): 805–30.

Yahav, Amit. "Time, Duration, and Defoe's Novels." *Partial Answers* 6, no. 1 (2008): 33–56.

———. "The Sense of Rhythm: Nationalism, Sympathy, and the English Elocutionists." *The Eighteenth Century* 52, no. 2 (2011): 173–92.

Zerubavel, Eviatar. *Hidden Rhythms: Schedules and Calendars in Social Life*. Chicago: University of Chicago Press, 1981.

Zimmerman, Everett. *The Boundaries of Fiction: History and the Eighteenth-Century British Novel*. Ithaca: Cornell University Press, 1996.

# Index


Abercrombie, David, 91, 170
Abrams, M. H., 10
Absorption, 2, 46–47, 108, 161–62
Action, 7, 14, 17–18, 19, 50–52, 112, 139–41; in Austen, 79–81; in Hutcheson, 58–64; in Richardson, 55–56, 64–65, 69–70, 73, 75, 77–78; in Sterne, 111
Addison, Joseph, 5, 16–17, 24–25, 38–45, 47–48, 50–52, 55, 117–18, 134, 161–62. Works: *The Spectator*: *no. 10*, 41; *no. 94*, 42–44; *no. 225*, 47–48, 50; *no. 287*, 41; *no. 418*, 41–42; *no. 476*, 38–40, 44
Aesthetics, 3–5, 15, 17, 19, 20, 116–47, 162; and Addison, 38, 40–43, 55; and Diderot, 45–46, 161; and Hume, 35, 48, 160
Agamben, Giorgio, 6, 158
Alkon, Paul, 153, 154
Anderson, Benedict, 153
Associationist conceptions, 3–4, 8, 16, 24, 33–34; in Addison, 38; in Austen, 147; in Radcliffe, 19, 117–19, 142, 145; in Richardson, 52, 55, 76–77; in Smith, 133, 136; in Sterne, 106–7, 172. *See also* succession of ideas
Attention, 27–28, 116, 135–36, 138, 145
Austen, Jane, 6, 79–81, 113–14, 145–47, 148

Baier, Annette, 159, 160
Bakhtin, Mikhail, 6–7, 9–10, 13, 44, 119, 156
Barker-Benfield, G. J., 85, 155
Beats. *See* Pulses
Bergson, Henri, 4–5, 7
Bildungsroman, 180
Blindness, 30–31,124–27; Molyneux's Problem, 176
Bonds, Mark Evan, 133, 178
Brissenden, R. F., 154–55
Brown, Marshall, 130, 177
Brown, Tony C., 59–60
Burke, Edmund, 5, 25, 40, 49; *A Philosophical Enquiry into the Origin of Our Ideas of the Sublime and Beautiful*, 19, 118–28, 130–31, 133–34, 175–76

Calendar, 30, 71–72, 166–67
Carey, Daniel, 164, 165
Carlson, Julia S., 10, 169–72
Castle, Terry, 137, 142, 179
Chandler, James, 6–7, 57, 63, 151–52, 155
Character, 12–13, 143; characterization, 6, 17, 19, 133, 136–42, 145, 147
Chartier, Roger, 162
Chronology, 7, 9, 11–12, 107, 111–12, 158, 161, 181–82
Chronometry, 2–3, 7, 9–12, 35–37, 41, 45, 48, 89, 112, 149–52; chronometric consciousness, 5–6, 8–10, 12, 19, 20, 40. *See also* measuring time; quantitative approaches to time
Chronotopes, 6–7, 10, 71–72, 156; chronometric chronotope, 9, 45, 153–54; romance chronotope, 13, 43–46; sensibility chronotope, 1–21, 43–46, 48–49, 76, 81, 98, 112–13, 115, 117, 119, 141–42, 144, 147–52, 180; 24/7 chronotope, 149
Clocks, 30–31, 99–100, 149–52
Clune, Michael W., 10, 156–57
Composition, 6, 14–16, 38, 40, 48, 133, 136, 139, 144–47, 157, 158, 180; compositional activity, 16, 19, 20, 23, 137, 148–52; compositional principles, 16–17, 34, 35, 37, 43, 134. *See also* form
Composure, 6, 138–39, 146. *See also* recollection
Consciousness, 2–4, 7, 14–17, 19, 117–19, 135, 149; in Locke, 26, 159; in Radcliffe, 130–31, 133, 137, 141–47; in Richardson, 52–53, 64, 76, 78–79

Conversation, 13, 17, 47–48, 101–2, 107–8
Crary, Jonathan, 19–20, 149

Dames, Nicholas, 146–47, 180
Darwall, Stephen, 60, 165
Defamiliarization, 45; Viktor Shklovsky, 161
Defoe, Daniel: *Robinson Crusoe*, 1–2, 8, 13–14, 21, 153
Deleuze, Gilles, 16, 35, 160
Deliberation, 51–52, 63, 65, 66, 70, 72, 80
De Marchi, Neil, 136
Diderot, Denis, 5, 16–17, 24–25, 127. Works: "Letter on the Blind," 125, 176–77; "In Praise of Richardson," 44–49, 53, 162
Digression, 13, 17, 18, 38, 44–45, 77–78, 80, 105, 108, 110–11
Discourse, 18, 84, 100–103, 109–11, 115, 136, 167–68, 174
Duncan, Ian, 11, 161, 174–75, 179
Dusk, 116–18, 128, 130–31

Elfenbein, Andrew, 95–96
Elocution, 5, 18, 86, 90–98, 105–6, 124, 128, 169–71; John Herries, 90, 93–95, 171; Thomas Sheridan, 90, 93–95, 171; Joshua Steele, 18, 90–93, 95–98, 102–3, 170
Emblematic time, 43, 50–51, 71–72, 161, 162
Embodied experience, 2, 15, 18, 48, 79–81, 86–91, 94–97, 156–57. *See also* sensation
Empiricist conceptions, 11–12, 16, 23–24, 38, 53, 120, 124, 180
Ermarth, Elizabeth Deeds, 11–12, 161, 162
Evening topos, 116, 128. *See also* nocturnal sensibility

Fellow-feeling. *See* sympathy
Ferguson, Frances, 137
Festa, Lynn, 155, 173
Finitude, 49, 109–10, 148–49
Foresight. *See* future projection
Form, 6–9, 11–19, 40, 42–43, 47, 70, 71, 102, 103, 108, 110, 118, 123–24, 136, 148, 152, 156–57, 162; gothic form, 119, 130, 144; linguistic form, 90–91, 94, 97, 171; lyric form, 116; musical form, 133–34, 140, 142; novel form, 12, 17, 52–54, 106, 147, 161, 163; rhythmic form, 85, 87–88, 104, 106, 111. *See also* composition
Foucault, Michel, 75–76, 153

Fried, Michael, 161–62
Future, 37, 39, 50, 75–76, 78; future projection, 40, 52, 53, 55, 68–70, 79, 138. *See also* prospection

Genette, Gérard, 77, 111–12, 114, 167–68, 174
Goodman, Kevis, 159
Gordon, Scott Paul, 155, 164
Gothic, 11, 19, 20, 72, 119, 130, 130–33, 142–49, 177
Gould, Glenn, 86–87, 105

Hartley, David, 3–4, 107
Heidegger, Martin, 4–5
Historicism, 11–12, 71; historical consciousness, 12, 19, 76, 143
Hogarth, William, 40, 122, 132, 175
Huhn, Tom, 122
Hume, David, 4–5, 16, 24, 25, 38, 48, 57, 63, 107, 122–23, 133–34, 155, 172. Works: *An Abstract of a Book Lately Published Entitled* A Treatise of Human Nature, 160; *A Treatise of Human Nature*, 33–37, 159
Husserl, Edmund, 4–5
Hutcheson, Francis, 5, 17–18, 49, 51–52, 56–68, 71, 76, 79, 80. Works: *An Essay on the Nature and Conduct of the Passions and Affections, with Illustrations on the Moral Sense*, 58, 61–64, 165; *An Inquiry into the Original of Our Ideas of Beauty and Virtue in Two Treatises*, 58–62, 165

Ideas, 25–27, 33, 88, 98–101. *See also* succession of ideas
Images, 125–27, 151–52
Impressions, 33–35
Infinite, 49, 61, 62, 64, 120–21, 149
Instants, 24, 25, 32–33, 89, 117, 119–20, 122–23
Intensities, 23–24, 31, 44, 45

Judgment, 3, 5, 17–20, 50–52, 79–80, 138; in Hutcheson, 60, 63–64, 164; in Richardson, 54–56, 64–66, 71–73, 78
Judgment of taste, 117–18

Kassler, Jamie C., 88, 169
Kermode, Frank, 21
Keymer, Thomas, 53, 110–11, 163, 173
Kinkead-Weekes, Mark, 53, 163
Kramnick, Jonathan, 64–65, 159

Lamb, Jonathan, 106–7, 158, 172, 173
Language, 15, 83, 86, 98, 106, 108, 124–28. *See also* discourse
Leisure, 8, 10, 13, 16, 40–41, 43, 46, 110, 150, 162
Lines, 40, 117–18, 122, 131–33
Locke, John: *An Essay Concerning Human Understanding*, 3–5, 16, 24–33, 48, 117, 148, 151; and Addison, 38, 42, 45; and Burke, 119, 122–24, 127, 176; and elocutionists, 91, 94; and Hume, 34–35; and musicology, 86, 88–89; and Smith, 134; and Sterne, 98–101, 172
Luhmann, Niklas, 67, 162, 166, 167
Lynch, Deidre Shauna, 8–9, 41, 79, 145, 155
Lyric, 10, 116–17, 128–30

Mack, Ruth, 143, 180
Macpherson, Sandra, 65, 139
Marclay, Christian: *The Clock*, 149–52, 181
Marshall, David, 97, 171–72
McKeon, Michael, 11–14, 18, 48, 111, 141
Measuring time, 9, 28–31, 35–37, 44, 112; in music, 89–90. *See also* calendars; chronometry; clocks
Mediation, 14–15, 18, 26–27, 159, 176
Memory, 27–28, 37, 39, 137–39, 144; and anticipation, 3–4, 12, 52, 118, 133–37, 151. *See also* recollection
Meter, 97–98, 104, 172
Miller, Christopher R., 10, 116–17, 129, 177
Mitchell, W. J. T., 123
Molesworth, Jesse, 77, 79, 167
Moment of good, 51, 60–64, 66–69, 71
Moments, 10, 13, 17, 24, 33–35, 61, 116–17, 138, 151; in Addison, 43; in Austen, 113–14; in Burke, 119–28, 133; in Diderot, 47; in Hume, 35, 160; in Richardson, 73–75, 78; in Sterne, 18, 82, 106–7, 110–11. *See also* instants; writing to the moment
Montage, 151–52, 181–82
Moral sense, 5, 17, 51, 57–60, 63, 66, 71, 164, 165
Moral sentiments, 17–19, 57, 63, 65–67
Moretti, Franco, 41, 180
Mullan, John, 106, 163–64
Music, 16, 18–19, 23, 84–90, 117–18, 148, 151, 178, 180; in Austen, 145–47; in elocution treatises, 95, 98, 171; in Hume, 35–36; in narratology, 114–15; in Radcliffe, 129, 133, 138–42; in Smith, 84–85, 133–36; in Sterne, 83, 104–7, 111, 172
Musicology, 5, 18, 86–91, 94, 97–98, 103, 133

Narration, 17, 18, 109–15
Narrative duration, 98, 110–15
Narrative length, 17, 52–56
Narratology, 77, 110, 112, 114, 158
Nazar, Hina, 57, 63, 65, 155
Nocturnal sensibility, 116, 128–33. *See also* evening topos
North, Roger, 18, 87–89

Parker, Patricia, 13, 44, 111
Performance, 95–97, 105–6
Periodicity, 25–26, 29, 41, 88
Phenomenology, 4–5, 16, 18, 23, 78, 117, 168
Pinch, Adela, 137, 178–79
Plot, 7, 14–15, 17–18, 52, 77–81, 111–12, 114–15, 140–41, 144; emplotment, 12, 15–16, 77, 180, 182; episodic, 16; suspense, 77–78
Poetics: antipictorialist poetics, 123–24, 126
Poetry, 129. *See also* lyric; Romantic poetry
Poulet, Georges, 23–24
Present, 72, 75, 108; present moment, 51, 54–55, 64; present tense, 53–54
Price, Leah, 54
Print, 3, 18, 40–41, 48, 90, 95–96, 142, 149, 152, 162
Prospection, 68–69, 71, 73, 77, 79, 80. *See also* future
Pulses, 84, 87–97, 115; in language, 93, 105; pulsating prose, 86, 97–110

Qualitative dimensions of time, 2–5, 9, 12, 14, 16–21, 24–25, 76, 93, 148–49, 158, 166; in Addison, 40, 42–43; in Diderot, 45–47; in Hume, 34–37; in Hutcheson, 52; in Locke, 31, 33; in Marclay, 150–52; in Smith, 134–35; in Sterne, 83–84, 99, 109–10, 111–12, 114–15. *See also* intensities
Quantitative approach to time, 25, 31, 33, 36, 42–43, 46, 49, 75; extensities and length, 29–31, 37. *See also* chronometry; measuring time

Radcliffe, Ann, 5, 13–15, 19, 20, 40, 49, 142–45, 174–75, 178, 180. Works: *The Italian*, 140; *The Mysteries of Udolpho*, 118–19, 128–33, 136–42

Reading as temporal mediation, 5, 8–10, 13–17, 20, 168; in Addison, 38, 40–43; in Diderot, 44, 47–49, 161–62; in Richardson, 52–56, 77–81; in Steele, 96–97; in Sterne, 102–3, 109–10
Recollection, 7, 9, 137–39, 142–47. *See also* composure; memory
Rhythm, 20; in conversation, 102; in elocution, 90–97; in language, 12, 18, 84, 86, 98, 171; in montage, 151–52; in music, 84, 86–90, 134–35; in poetry, 10; in print, 95–96, 102; in prose, 98–99, 104, 106–7, 109–12; in sympathy, 17–19, 82–115; rhythmic continuity, 87
Richardson, Alan, 120
Richardson, Samuel, 5, 6, 13, 17–18, 49, 51–56, 64–81, 138, 147; Diderot on, 44–47. Works: *Clarissa*, 65, 68–77, 139; *Letters*, 53–56, 77; *Pamela*, 66–68, 71–72, 166–67
Ricoeur, Paul, 14–16, 20, 157
Riskin, Jessica, 44, 155
Romance, 9–14, 18–20, 161; in Addison, 43–44; in Diderot, 44–46; in Radcliffe, 118–19, 142–47, 175; in Richardson, 78; in Sterne, 83, 110–11
Romantic poetry, 9–11, 130, 158
Romantic time, 116, 129, 156–57
Rousseau, Jean-Jacques, 134, 171, 178
Rowland, Ann Wierda, 130

Sedgwick, Eve Kosofsky, 132, 133
Sensation, 3–5, 19, 23, 29, 120, 126–28, 158; sensory evidence, 26, 30–31. *See also* embodied experience; empiricist conceptions
Sensationist conceptions, 16, 19, 24, 123, 130, 147, 152
Sherman, Stuart, 7–9, 36, 41, 53, 76, 103, 156
Silver, Sean, 38, 144, 180
Smith, Adam, 5, 17, 49, 57, 110, 163. Works: "Of the Nature of that Imitation Which Takes Place in What are Called the Imitative Arts," 19, 118, 133–36, 138, 178; *The Theory of Moral Sentiments*, 84–85, 97, 171–72
Social dimensions of temporality, 2–3, 5, 8, 47–49, 162
Soni, Vivasvan, 17, 56–57, 73, 79, 163, 168
Sterne, Laurence, 5, 6, 13–15, 18, 25, 49, 86, 115, 124, 128, 147, 158, 172, 173, 176. Works: "Letter to Mrs. Vesey," 82–84, 86, 168; *Tristram Shandy*, 15, 77, 86, 98–112
Stewart, Garrett, 114–15, 151–52, 174, 181–82
Succession of ideas, 24, 86, 89, 117, 119, 127, 133; in Addison, 38–40, 42–44; in Diderot, 48; in Hume 34–37; in Locke, 3, 25–33; in Marclay, 151; in Smith, 133, 134–35; in Steele, 91; in Sterne, 98, 100–101; succession without ideas, 120–23, 134
Succession of sensations, 120, 126, 128–29
Succession of sounds, 134–36
Sympathy, 5, 14, 17–19, 51, 57, 72–73, 82–115; in elocutionist treatises, 91, 93–95, 97; in Sterne, 84, 99, 106–7, 109–11
Synchronization, 18, 94–95, 102, 106, 108, 110
Synchrony and succession, 3–4

Thompson, E. P., 153
Truffaut, François: *Fahrenheit 451*, 132–33
Turetzky, Philip, 26, 159
Turrell, James, 116–18, 130–31
Twilight. *See* dusk

Utilitarian approaches, 8, 18, 41, 51–52, 60–61, 63–64

Van Sant, Ann Jessie, 155, 158
Vision, 120–25, 128–32. *See also* blindness

Walpole, Horace, 19, 119, 142–44, 179, 180
Warner, William B., 75
Watt, Ian, 53, 64, 79
Wordsworth, William, 9, 97, 138
Writing to the moment, 18, 52–56

Zerubavel, Eviatar, 153–54

# *Acknowledgments*

It is my great luck to have had remarkable teachers whose inspiration lured me to the otherwise unlikely destiny of scholar of eighteenth-century English literature. At Tel Aviv University, Hanna Wirth-Nesher's magical readings of Virginia Woolf and Henry James first suggested the possibility of immersing myself in literature. And at Johns Hopkins University, Frances Ferguson's dazzling classes on novels, critical theory, and legal history won me over to the eighteenth century. Frances, together with Amanda Anderson and the late Allen Grossman, taught me how to think, read, and write in ways that I never imagined I could. Their teachings shaped my understanding well beyond my professional inquiries. After graduate school, Frances Ferguson followed up with unflagging support as I embarked on the research project that culminated in this book. Mere words of gratitude can hardly suffice to acknowledge such generosity; with her example in mind, I do my best to stand by the students I mentor.

At Johns Hopkins University I was also lucky to meet Irene Tucker, Ruth Mack, and Hina Nazar. Their friendship and professional models inspired me through graduate school and beyond. Vered Lev Kenaan, Eran Shalev, and Zur Shalev provided much needed camaraderie through my years at the University of Haifa. Itamar Levi remains an inexhaustible resource for conversations about art, literature, philosophy, and psychoanalysis. At the University of Minnesota the welcoming warmth of Elaine Auyoung, Siobhan Craig, Lianna Farber, Brian Goldberg, Qadri Ismail, Pamela Leszczynski, Ellen Messer-Davidoff, Katherine Scheil, Andrew Scheil, and J. B. Shank makes even a long Minnesotan winter cheerful. And having Andrew Elfenbein as a senior colleague makes all the difference in the world. I owe him special thanks for reading the full manuscript at its later stages and for his candid advice on writing and revising.

Elaine Auyoung, Tony Brown, Sarah Gilead, Brian Goldberg, Frances

Ferguson, and Irene Tucker provided thoughtful and erudite comments on various chapters. Daniel Brewer, Juliette Cherbuliez, Mary Franklin-Brown, Jennifer Row, and J. B. Shank—my colleagues in the Temporalities group at the Consortium for the Study of the Premodern World—ensured stimulating discussions of all things temporal and offered helpful remarks on the introduction and the first chapter. Two anonymous reviewers for the University of Pennsylvania Press delivered remarkably insightful assessments of the manuscript. Their stunning readings enabled me to add two crucial reconceptualizations during the final preparation of the book. I am grateful to them as well as to Jerome Singerman, whose wisdom must have been responsible for this felicitous match. Amy Fairgrieve proofread and fact-checked the entire manuscript with extraordinary diligence and intelligence.

A portion of my analysis of Sterne's rhythmic narration has been published in *PMLA* (128.4 [2013]: 872–887), an offshoot of my discussion of the elocutionists has been published in *The Eighteenth Century: Theory and Interpretation* (52.2 [2011]: 173–92), and my preliminary thoughts on qualitative duration have been published in *Partial Answers* (6.1 [2008]: 33–56). Many thanks to the editors and anonymous reviewers for these journals, who provided helpful comments and gave me a chance to test in print some of my research in progress. I also owe gratitude to the Newberry Library, the Clark Center for Seventeenth- and Eighteenth-Century Studies, and the Lewis Walpole Library for supporting this project with fellowships, as well as to the University of Haifa for a sabbatical leave during which I was an affiliate fellow at the Center for Cultural Analysis at Rutgers University. Lynn Festa, Jonathan Kramnick, Meredith McGill, and Henry Turner, welcomed me at Rutgers; they, along with my fellows in that year's CCA seminar, made residence at Rutgers a delightful intellectual treat. Course releases from the University of Minnesota's English Department and a fellowship from the Institute for Advanced Study at the University of Minnesota enabled me to complete the manuscript. Special thanks to Jennifer Gunn and Susannah Smith for tolerating my hermit preferences at the IAS, and to my cohort of fellows in the fall of 2015, whose weekly conversations over lunch kept my communicative abilities in order.

I am grateful to all the people mentioned in these acknowledgments. I also wish to thank the many others—peers, colleagues, students, neighbors, and friends—who have made the institutions and cities I resided in or passed through the stimulating places that they are. My greatest debts are owed to

Aviva and the late Yossi Yahav, who guaranteed emotional and material support throughout my life, to Yaniv Garama who sustained me through the years of working on this project with contact improvisation, shiatsu, and good laughs, and to Ye'ela and Mikael Garama-Yahav, who give "huggies" without which I simply cannot do. I dedicate this book with much love to the luminous memory of Yossi and to the vibrant presence of Ye'ela and Mikael.

CPSIA information can be obtained
at www.ICGtesting.com
Printed in the USA
BVHW03*1700260318
511366BV00001B/1/P

9 780812 250176